Video Based Machine Learning for Traffic Intersections

Video Based Machine Learning for Traffic Intersections describes the development of computer vision and machine learning–based applications for intelligent transportation systems (ITS) and the challenges encountered during their deployment. This book presents several novel approaches, including a two-stream convolutional network architecture for vehicle detection, tracking, and near-miss detection; an unsupervised approach to detect near-misses in fisheye intersection videos using a deep learning model combined with a camera calibration and spline-based mapping method; and algorithms that utilize video analysis and signal timing data to accurately detect and categorize events based on the phase and type of conflict in pedestrian–vehicle and vehicle–vehicle interactions.

The book makes use of a real-time trajectory prediction approach, combined with aligned Google Maps information, to estimate vehicle travel time across multiple intersections. Novel visualization software, designed by the authors to serve traffic practitioners, is used to analyze the efficiency and safety of intersections. The software offers two modes: a streaming mode and a historical mode, both of which are useful to traffic engineers who need to quickly analyze trajectories to better understand traffic behavior at an intersection.

Overall, this book presents a comprehensive overview of the application of computer vision and machine learning to solve transportation-related problems. *Video Based Machine Learning for Traffic Intersections* demonstrates how these techniques can be used to improve safety, efficiency, and traffic flow, as well as identify potential conflicts and issues before they occur. The range of novel approaches and techniques presented offers a glimpse of the exciting possibilities that lie ahead for ITS research and development.

Key Features:

- Describes the development and challenges associated with Intelligent Transportation Systems (ITS)
- Provides novel visualization software designed to serve traffic practitioners in analyzing the efficiency and safety of an intersection
- Has the potential to proactively identify potential conflict situations and develop an early warning system for real-time vehicle–vehicle and pedestrian–vehicle conflicts

Video Based Machine Learning for Traffic Intersections

Tania Banerjee, Xiaohui Huang,
Aotian Wu, Ke Chen, Anand Rangarajan
and Sanjay Ranka

CRC Press is an imprint of the
Taylor & Francis Group, an **informa** business

Designed cover image: © Shutterstock ID 260656910, Photo Contributor Joshua Davenport

First edition published 2024
by CRC Press
6000 Broken Sound Parkway NW, Suite 300, Boca Raton, FL 33487-2742

and by CRC Press
4 Park Square, Milton Park, Abingdon, Oxon, OX14 4RN

CRC Press is an imprint of Taylor & Francis Group, LLC

ISBN: 978-1-032-54226-3 (hbk)
ISBN: 978-1-032-56517-0 (pbk)
ISBN: 978-1-003-43117-6 (ebk)

DOI: 10.1201/9781003431176

Typeset in TexGyreTermesX-Regular font
by KnowledgeWorks Global Ltd.

Dedication

I dedicate this book to my loving family, friends and mentors who supported me every step of the way, and to all the readers who have picked up this book.

–Xiaohui

I dedicate this book to future professionals who will further advance the field of intelligent transportation systems and shape the transportation systems of tomorrow.

–Aotian

I dedicate this book to my beloved son, who has been my source of inspiration and motivation, and to my dear parents, whose unwavering love and support have been a constant throughout my journey.

–Tania

I dedicate this book to my wife, Deepa.

–Sanjay

Contents

Disclaimer

This book is intended for pedagogical purposes and the authors have taken care to provide details based on their experience. However, the authors do not guarantee that the material in this book is accurate or will be effective in practice, nor are they responsible for any statement, material, or formula that may have negative consequences, injuries, death, etc. to the readers or the users of the book.

List of Figures

List of Tables

Authors

Tania Banerjee is a research assistant scientist in Computer and Information Science and Engineering at the University of Florida. She earned her PhD in Computer Science from the University of Florida, Gainesville, Florida, in 2012. She completed her M Sc in Mathematics at the Indian Institute of Technology, Kharagpur. Her research interests are video analytics, intelligent transportation, data compression, and high-performance computing.

Xiaohui Huang earned her PhD in the Department of Computer & Information Science & Engineering, University of Florida, Gainesville, Florida, in 2020. Her research interests include machine learning, computer vision, and intelligent transportation systems.

Aotian Wu is at present a PhD student in the Department of Computer & Information Science & Engineering, University of Florida, Gainesville, Florida. Her research interests are machine learning, computer vision, and intelligent transportation systems.

Ke Chen is at present a PhD student in the Department of Computer & Information Science & Engineering, University of Florida, Gainesville, Florida. His research interests are machine learning, computer architecture, operating systems and algorithms, and data structures.

Anand Rangarajan is a Professor in the Department of CISE, University of Florida, Gainesville, Florida. His research interests are machine learning, computer vision, medical and hyperspectral imaging, and the science of consciousness.

Sanjay Ranka is a Distinguished Professor in the Department of Computer Information Science and Engineering at University of Florida. His current research interests are high-performance computing and big data science with a focus on applications in CFD, healthcare, and transportation. He has coauthored four books, 290+ journals, and refereed several conference articles. He is a Fellow of the IEEE and AAAS. He is Associate Editor-in-Chief of the *Journal of Parallel and Distributed Computing* and Associate Editor for *ACM Computing Surveys, Applied Sciences, Applied Intelligence, IEEE/ACM Transactions on Computational Biology and Bioinformatics.*

1 Introduction

1.1 MOTIVATION

Rapid urbanization worldwide, the growing volume of vehicular traffic, and the increasing complexity of roadway networks have led to congestion, traffic jams, and traffic incidents [1, 2], which negatively affect productivity [3], the well-being of the society [4], and the environment [5]. Therefore, keeping traffic flowing smoothly and safely is essential for traffic engineers.

Significant advances in electronics, sensing, computing, data storage, and communications technologies in intelligent transportation systems (ITS) [6, 7] have led to some commonly seen aspects of ITS [8] which include the following:

- Microprocessor-based traffic signals "coordinated" across intersections to optimize traffic flow across a corridor. These microprocessors communicate with embedded road sensors that relay vehicle detection back to the microprocessors.
- Collection and storage of high-resolution (10 Hz) induction loop detector actuations, coupled with detailed signal state information along with visualization of derived metrics called Automated Traffic Signal Performance Measures (ATSPM).
- Video cameras (with computer vision algorithms) at intersections to detect offenders running the stoplight, vehicle tracking, queue length estimation, etc.

This book presents algorithms and methods for using video analytics to process traffic data. Edge-based real-time machine learning (ML) techniques and video stream processing have several advantages: (1) There is no need to store extensive amounts of video (a few minutes typically suffice for edge-based processing), thus addressing concerns of public agencies who do not want person-identifiable information to be stored for reasons of citizen privacy and legality. (2) The processing of the video stream at the edge will allow low-bandwidth communication using wireline and wireless networks to a central system such as a cloud, resulting in a compressed and holistic picture of the entire city. (3) Real-time processing makes it possible to develop a range of new transportation applications at the intersection, street, and system levels that were previously impossible. This has a significant impact on safety and mobility, bringing about novel opportunities that were not available until now.

The existing monitoring systems and decision-making for this purpose have several limitations:

- Current sensors have limited capability: Vehicle loop detectors that have traditionally been deployed at intersections to detect the passage of vehicles are error-prone; have high deployment and maintenance costs; can only

DOI: 10.1201/9781003431176-1

Table 1.1
Raw Event Logs from Signal Controllers

SignalID	Timestamp	EventCode	EventParam
1490	2018-08-01 00:00:00.000100	82	3
1490	2018-08-01 00:00:00.000300	82	8
1490	2018-08-01 00:00:00.000300	0	2
1490	2018-08-01 00:00:00.000300	0	6
1490	2018-08-01 00:00:00.000300	46	1
1490	2018-08-01 00:00:00.000300	46	2
1490	2018-08-01 00:00:00.000300	46	3

Most modern controllers generate these data at a frequency of 10 Hz.

measure the absence or presence of vehicles passing above them; and are not always helpful in observing the movements of pedestrians and scooters. When the sensors are not accurate or timely, an adaptive strategy will not be effective. Video detection can improve accuracy and timeliness in detecting vehicles, pedestrians, bicyclists, etc.

- Current software systems for traffic monitoring need to be more cohesive and suitable for real-time decision-making: Transportation professionals have to deal with an abundance of disconnected data that is spread across different systems. Existing intersection control systems do not provide reports on a real-time basis (based on the vendor), and these are given at coarse levels of granularity (for example, traffic movement counts by the hour) limiting their use and ability to make real-time changes to adapt to dynamically changing conditions. Current approaches are not readily scalable because of constraints of cost, bandwidth, and lack of integration.

In Section 1.2, we will describe the data sources from which we collect and store loop recorder data and video data. Section 1.3 summarizes the rest of the chapters in the book.

1.2 DATA SOURCES

1.2.1 INTERSECTION CONTROLLER LOGS

Signal controllers, based on the latest Advanced Transportation Controller (ATC)[1] standards, are capable of recording intersection events (e.g., signal changes and vehicle arrival and departure events) at a high data rate (10 Hz). The different attributes in the data include intersection identifier, timestamp, and EventCode and EventParam. The accompanying metadata describes what different EventCode and EventParam

[1] https://www.ite.org/technical-resources/standards/atc-controller/

indicate. For instance, EventCode 82 denotes vehicle arrival, and the corresponding EventParam represents the detector channel that captured the event. The other necessary metadata is the detector channel to lane/phase mapping, which helps to identify the lane corresponding to a specific detector channel. Different performance measures of interest, such as arrivals on red, arrivals on green, or demand-based split failures, can be derived on a granular level (cycle-by-cycle) using these data.

1.2.2 VIDEO DATA

We process videos captured by fisheye cameras installed at intersections. A fisheye (bell) camera has an ultrawide angle lens, resulting in a wide panoramic view of the nonrectilinear image. Acquiring locations with a fisheye camera is advantageous because it can obtain a complete picture of the entire intersection.

The fisheye intersection videos are more challenging than videos collected by surveillance cameras for reasons including fisheye distortion, multiple object types (pedestrians and vehicles), and diverse lighting conditions. We annotated the spatial location (bounding boxes) and temporal location (frames) for each object and their vehicle class in videos for each intersection to generate ground truth for object detection, tracking, and near-miss detection.

1.3 CHAPTER ORGANIZATION

The chapters are organized as follows. In Chapter 2, we propose an integrated two-stream convolutional network architecture that performs real-time detection, tracking, and near-miss detection of road users in traffic video data. The two-stream model consists of a spatial stream network for object detection and a temporal stream network to leverage motion features for multiple object tracking. We detect near-misses by incorporating appearance features and motion features from these two networks. Further, we demonstrate that our approaches can be executed in real time and at a frame rate higher than the video frame rate on various videos.

In Chapter 3, we introduce trajectory clustering and anomaly detection algorithms. We develop real-time or near real-time algorithms for detecting near-misses for intersection video collected using fisheye cameras. We propose a novel method consisting of the following steps: (1) extracting objects and multiple object tracking features using convolutional neural networks; (2) densely mapping object coordinates to an overhead map; and (3) learning to detect near-misses by new distance measures and temporal motion. The experiments demonstrate the effectiveness of our approach with a real-time performance at 40 fps and high specificity.

Chapter 4 presents an end-to-end software pipeline for processing traffic videos and running a safety analysis based on surrogate safety measures. As a part of road safety initiatives, surrogate road safety approaches have gained popularity due to the rapid advancement of video collection and processing technologies. We developed algorithms and software to determine trajectory movement and phases that, when combined with signal timing data, enable us to perform accurate event detection and categorization regarding the type of conflict for both pedestrian–vehicle and vehicle–vehicle interactions. Using this information, we introduce a new surrogate safety measure, "severe event," which is quantified by multiple existing metrics such as time-to-collision (TTC) and post-encroachment time (PET) as recorded in the event, deceleration, and speed. We present an efficient multistage event-filtering approach followed by a multi-attribute decision tree algorithm that prunes the extensive set of conflicting interactions to a robust set of severe events. The above pipeline was used to process traffic videos from several intersections in multiple cities to measure and compare pedestrian and vehicle safety. Detailed experimental results are presented to demonstrate the effectiveness of this pipeline.

Chapter 5 illustrates cutting-edge methods by which conflict hotspots can be detected in various situations and conditions. Both pedestrian–vehicle and vehicle–vehicle conflict hotspots can be discovered, and we present an original technique for including more information in the graphs with shapes. Conflict hotspot detection, volume hotspot detection, and intersection-service evaluation allow us to comprehensively understand the safety and performance issues and test countermeasures. The selection of appropriate countermeasures is demonstrated by extensive analysis and discussion of two intersections in Gainesville, Florida. Just as important is the evaluation of the efficacy of countermeasures. This chapter advocates for selection from a menu of countermeasures at the municipal level, with safety as the top priority. Performance is also considered, and we present a novel concept of a performance–safety trade-off at intersections.

In Chapter 6, we propose to perform trajectory prediction using surveillance camera images. As vehicle-to-infrastructure (V2I) technology enables low-latency wireless communication, warnings from our prediction algorithm can be sent to vehicles in real time. Our approach consists of an offline learning phase and an online prediction phase. The offline phase learns common motion patterns from clustering, finds prototype trajectories for each cluster, and updates the prediction model. The online phase predicts the future trajectories for incoming vehicles, assuming they follow one of the motion patterns learned from the offline phase. We adopted a long short-term memory encoder–decoder (LSTM-ED) model for trajectory prediction. We also explored using a curvilinear coordinate system (CCS) which utilizes the learned prototype and simplifies the trajectory representation. Our model is also able to handle noisy data and variable-length trajectories. Our proposed approach outperforms the baseline Gaussian process (GP) model and shows sufficient reliability when evaluated on collected intersection data.

In Chapter 7, we propose a methodology for travel time estimation of traffic flow, an important problem with critical implications for traffic congestion analysis. We developed techniques for using intersection videos to identify vehicle trajectories across multiple cameras and analyze corridor travel time. Our approach consists of (1) multi-object single-camera tracking, (2) vehicle reidentification among different cameras, (3) multi-object multi-camera tracking, and (4) travel time estimation. We evaluated the proposed framework on real intersections in Florida with pan and fisheye cameras. The experimental results demonstrate the viability and effectiveness of our method.

In Chapter 8, we present a visual analytics framework that traffic engineers may use to analyze the events and performance at an intersection. The tool ingests streaming videos collected from a fisheye camera, cleans the data, and runs analytics. The tool presented here has two modes–streaming and historical modes. The streaming mode may be used to analyze data close to real time with a latency set by the user. In the historical mode, the user can run a variety of trend analyses on historical data.

In Chapter 9, we summarize the contributions of the present work.

2 Detection, Tracking, and Classification

2.1 INTRODUCTION

The rapid changes in the growth of exploitable and, in many cases, open data can mitigate traffic congestion and improve safety. Despite significant advances in vehicle technology, traffic engineering practices, and analytics based on crash data, the number of traffic crashes and fatalities is still too many. Many drivers are frustrated due to prolonged (potentially preventable) intersection delays. Using video or light detection and ranging (LiDAR) processing, big data analytics, artificial intelligence (AI), and machine learning can profoundly improve the ability to address these challenges. Collecting and exploiting large datasets are familiar to the transportation sector. However, the confluence of ubiquitous digital devices and sensors, significantly lower hardware costs for computing and storage, enhanced sensing and communication technologies, and open-source analytics solutions have enabled novel applications. The latter may involve insights into otherwise unobserved patterns that positively influence individuals and society.

The technologies of AI and the Internet of Things (IoT) are ushering in a new promising era of "smart cities" where billions of people around the world can improve the quality of their lives in aspects of transportation, security, information, communications, etc. One example of data-centric AI solutions is computer vision technologies that enable vision-based intelligence for edge devices across multiple architectures. Sensor data from smart devices or video cameras can be analyzed immediately to provide real-time analysis for intelligent transportation systems (ITS). At traffic intersections, there is a greater volume of road users (pedestrians and vehicles), traffic movement, dynamic traffic events, near-accidents, etc. It is a critically important application to enable global monitoring of traffic flow, local analysis of road users, and automatic near-miss detection.

As a new technology, vision-based intelligence has many applications in traffic surveillance and management [9, 10, 11, 12, 13, 14]. Many research works have focused on traffic data acquisition with aerial videos [15, 16]; the aerial view provides better perspectives to cover a large area and focus resources for surveillance tasks. Unmanned aerial vehicles (UAVs) and omnidirectional cameras can acquire helpful aerial videos for traffic surveillance, especially at intersections, with a broader perspective of the traffic scene and the advantage of being mobile and spatiotemporal. A recent trend in vision-based intelligence is to apply computer vision technologies to these acquired intersection aerial videos [17, 18] and process them at the edge across multiple ITS architectures.

Object detection and multiple object tracking(MOT) are widely used applications in transportation, and real-time solutions are significant, especially for the emerging

DOI: 10.1201/9781003431176-2

area of big transportation data. A near-miss is an event that has the potential to develop into a collision between two vehicles or between a vehicle and a pedestrian or bicyclist. These events are important to monitor and analyze to prevent crashes in the future. They are also a proxy for potential timing and design issues at the intersection. Camera-monitored intersections produce video data in gigabytes per camera per day. Analyzing thousands of trajectories collected per hour at an intersection from different sources to identify near-miss events quickly becomes challenging, given the amount of data to be examined and the relatively rare occurrence of such events.

In this work, we investigate using traffic video data for near-miss detection. However, to the best of our knowledge, a unified system that performs real-time detection and tracking of road users and near-miss detection for aerial videos is not available. Therefore, we have collected video datasets and presented a real-time deep learning–based method to tackle these problems.

Generally, a vision-based surveillance tool for ITS should meet several requirements: (1) segment vehicles from their surroundings (including other road objects and the background) to detect all road objects (still or moving); (2) classify detected vehicles into categories: cars, buses, trucks, motorbikes, etc.; (3) extract spatial and temporal features (motion, velocity, and trajectory) to enable more specific tasks, including vehicle tracking, trajectory analysis, near-miss detection, anomaly detection, etc.; (4) function under various traffic conditions and lighting conditions; and (5) operate in real time. Over the decades, although increasing research on vision-based systems for traffic surveillance has been proposed, many of the criteria listed above still need to be met. Early solutions [19] do not identify individual vehicles as unique targets and progressively track their movements. Methods have been proposed to address individual vehicle detection and vehicle tracking problems [20, 21, 9] with tracking strategies and optical flow deployment. Compared to traditional hand-crafted features, deep learning methods [22, 23, 24, 25, 26, 27] in object detection have illustrated the robustness of specialization of the generic detector to a specific scene. Recently, automatic traffic accident detection has become an important topic. Before detecting accident events [28, 12, 29, 30, 31], one typical approach is to apply object detection or tracking methods using a histogram of flow gradient (HFG), hidden Markov model (HMM), or Gaussian mixture model (GMM). Other approaches [32, 33, 34, 35, 36, 37, 38, 39, 40] use low-level features (e.g., motion features) to demonstrate better robustness. Neural networks have also been employed for automatic accident detection [41, 42, 43, 44].

The overall pipeline of our method is depicted in Figure 2.1. The organization of this chapter is as follows. Section 2.2 describes the background of convolutional neural networks (CNN), object detection, and MOT methods. Section 2.3 describes our method's overall architecture, methodologies, and implementation. This is followed in Section 2.4 by introducing our traffic near-accident detection dataset (TNAD) and presenting a comprehensive evaluation of our approach and other state-of-the-art near-accident detection methods, both qualitatively and quantitatively. Section 2.5 summarizes our contributions and discusses future work's scope.

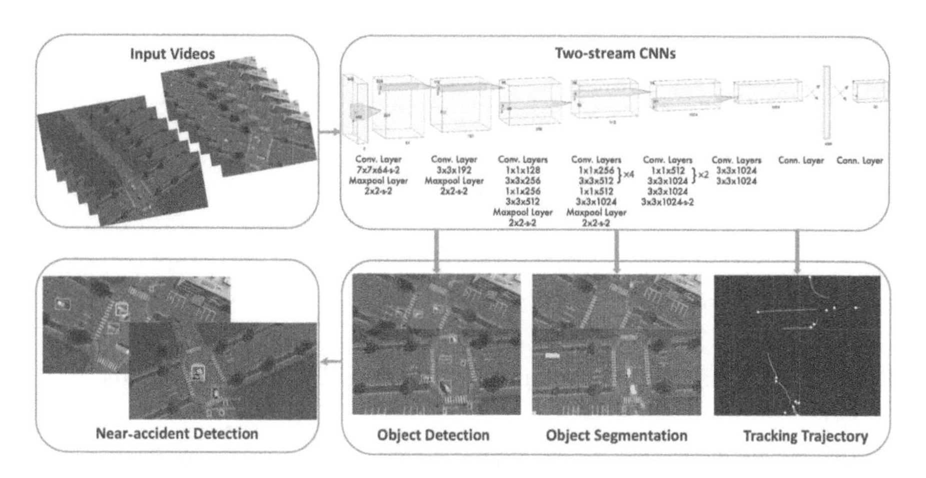

Figure 2.1 The overall pipeline of our two-stream convolutional neural networks for near-miss detection

2.2 COMPUTER VISION APPROACHES

2.2.1 CONVOLUTIONAL NEURAL NETWORKS

CNNs have shown strong capabilities in representing objects, thereby boosting the performance of numerous vision tasks, especially compared to traditional features [45]. A CNN is a class of deep neural networks which is widely applied in image analysis and computer vision. A standard CNN usually consists of both the input layer and the output layer, as well as multiple hidden layers (e.g., convolutional layers, fully connected layers, pooling layers), as shown in Figure 2.2. The input to a convolutional layer is an original image $\boldsymbol{X}$. We denote the feature map of the ith convolutional layer as $\boldsymbol{H}_i$, and $\boldsymbol{H}_0 = \boldsymbol{X}$. Then, $\boldsymbol{H}_i$ can be described as follow:

$$\boldsymbol{H}_i = f\left(\boldsymbol{H}_{i-1} \otimes \boldsymbol{W}_i + \boldsymbol{b}_i\right), \tag{2.1}$$

where $\boldsymbol{W}_i$ is the weight for the ith convolutional kernel for the $i-1$-th image or feature map and $\otimes$ is the convolution operation. The output of the convolution operation includes a bias, $\boldsymbol{b}_i$. Then, the feature map for the ith layer can be computed by applying a standard nonlinear activation function. We briefly describe a 32×32 RGB image with a simple ConvNet for CIFAR-10 image classification [46]:

- *Input layer*: The original 32×32 image with three color channels (R, G, B).
- *Convolutional layer*: Local operations in the image passed through nonlinear activation functions. If we use 12 filters, the size of the result is $(32 \times 32 \times 12)$.
- *Pooling layer*: A downsampling operation, resulting in a volume such as $(16 \times 16 \times 12)$.
- *Fully connected layer*: Output scores for each class, resulting in a size volume $(1 \times 1 \times 10)$ for ten classes.

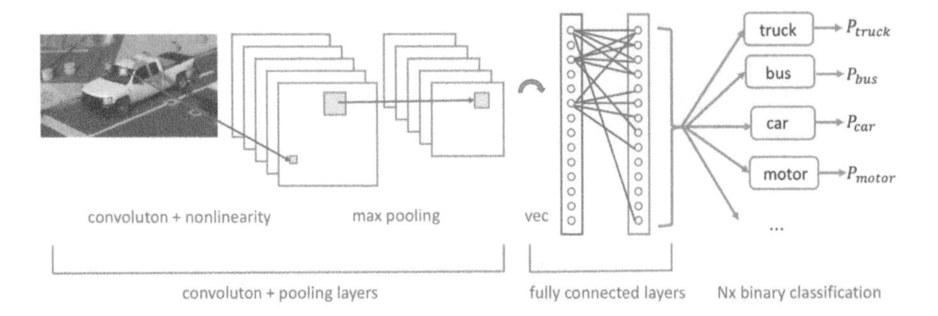

Figure 2.2 Architecture of convolutional neural networks for image classification

In this way, CNNs transform the original image into multiple high-level feature representations layer by layer, obtaining class-specific outputs or scores.

2.2.2 YOLO OBJECT DETECTION

The real-time You Only Look Once (YOLO) detector, proposed in Ref. [24], is an end-to-end state-of-the-art deep learning approach without using region proposals. The pipeline of YOLO [24] is relatively straightforward: Given an input image, YOLO [24] passes it through the neural network only once, as its name implies (You Only Look Once), and outputs the detected bounding boxes and class probabilities in prediction. Figure 2.4 demonstrates the detection model and system of YOLO [24]. YOLO [24] is orders of magnitude faster (45 frames per second) than other object detection approaches, which means it can process streaming video in real time. Compared to other systems, it also achieves a higher mean average precision. In this work, we leverage the extension of YOLO [24], Darknet-19, a classification model used as the basis of YOLOv2 [47]. Darknet-19 [47] consists of 19 convolutional layers and 5 max-pooling layers, where batch normalization is utilized to stabilize training, speed up convergence, and regularize the model [48].

2.2.3 SIMPLE ONLINE REAL-TIME TRACKING (SORT)

SORT [49] is a simple, popular, fast MOT algorithms. The core idea combines Kalman filtering [50] and frame-by-frame data association. The data association is implemented with the Hungarian method [51] by measuring the bounding box overlap. With this rudimentary combination, SORT [49] achieves a state-of-the-art performance compared to other online trackers. Moreover, due to its simplicity, SORT [49] can update at a rate of 260 Hz on a single machine, which is over 20 times faster than other state-of-the-art trackers.

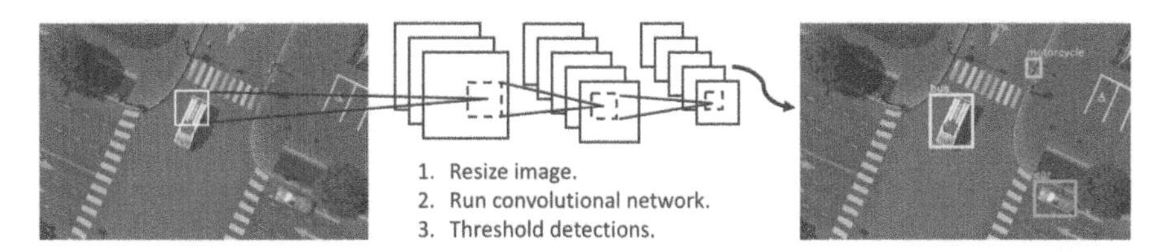

1. Resize image.
2. Run convolutional network.
3. Threshold detections.

Figure 2.3 Object detection pipeline of our spatial stream [24]: (1) resizes the input video frame, (2) runs convolutional network on the frame, and (3) thresholds the resulting detection by the model's confidence

2.2.4 DeepSORT

DeepSORT [52] is an extension of SORT [49]. DeepSORT integrates appearance information to improve the performance of SORT [49] by adding one pre-trained association metric. DeepSORT [52] helps solve many identity-switching problems in SORT [49], and it can track occluded objects in a longer term. The measurement-to-track association is established in visual appearance space during the online application, using nearest-neighbor queries.

2.3 TWO-STREAM ARCHITECTURE FOR NEAR-MISS DETECTION

This section presents our computer vision–based two-stream architecture for real-time near-miss detection. The architecture is primarily driven by real-time object detection and MOT. The goal of near-accident detection is to detect likely collision scenarios across video frames and report these near-miss records. Because videos have spatial and temporal components, we divide our framework into a two-stream architecture, as shown in Figure 2.3. The spatial aspect comprises individual frame appearance information about scenes and objects. The temporal element comprises motion information of objects. For the spatial stream convolutional neural network, we utilize a standard convolutional network designed for state-of-the-art object detection [24] to detect individual vehicles and mark near-miss regions at the single-frame level. The temporal stream network leverages object candidates from object detection CNNs and integrate their appearance information with a fast MOT method to extract motion features and compute trajectories. When two trajectories of individual objects intersect or come closer than a certain threshold (whose estimation is described below), we label the region covering the two entities as a high probability near-miss area. Finally, we take the average near-miss likelihood of both the spatial and temporal stream networks and report the near-miss record.

2.3.1 OBJECT DETECTION AND CLASSIFICATION

Each stream is implemented using a deep convolutional neural network in our framework. Near-accident scores are combined by averaging. Because our spatial stream

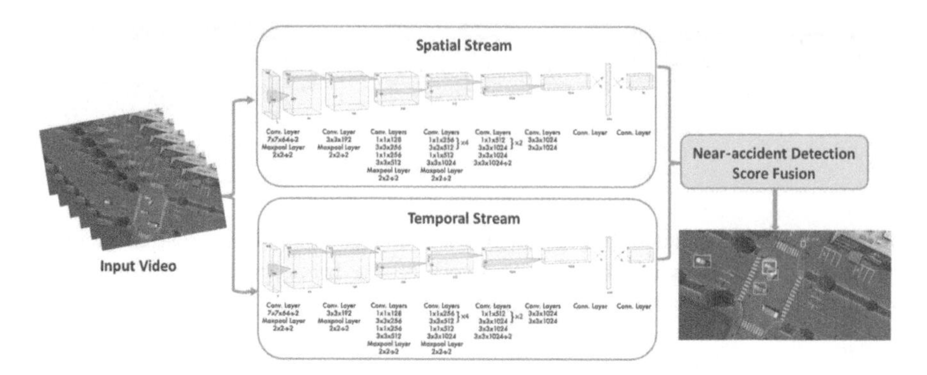

Figure 2.4 Two-stream convolutional neural networks architecture for near-miss detection

ConvNet is essentially an object detection architecture, we base it on recent advances in object detection – essentially the YOLO detector [24] – and pre-train the network from scratch on our dataset containing multiscale drone, fisheye, and simulation videos. As most of our videos have traffic scenes with vehicles and movement captured in a top-down view, we specify different vehicle classes such as motorbike, car, bus, and truck as object classes for training the detector. Additionally, near-misses or collisions can be detected from single still frames or stopped vehicles associated with an accident, even at the beginning of a video. Therefore, we train our detector to localize these likely near-miss scenarios. Since static appearance is a valuable cue, the spatial stream network performs object detection by only operating on individual video frames.

The spatial stream network regresses the bounding boxes and predicts the class probabilities associated with these boxes using a simple end-to-end convolutional network. It first splits the image into a $S \times S$ grid. For each grid cell, it does the following:

- Predicts B boundary boxes, and each box has a confidence score
- Detects one object only, regardless of the number of boxes B
- Predicts C conditional class probabilities (one per class for the likelihood of the object class)

For each bounding box, the CNN outputs a class probability and offset values for the bounding box. Then, it selects bounding boxes that have the class probability above a threshold value and uses them to locate the object within the image. In essence, each boundary box contains five elements: (x, y, w, h) and box confidence. The (x, y) are coordinates that represent the box's center relative to the grid cell's bounds. The (w, h) parameters are the width and height of the object. These elements are normalized such that x, y, w, and h lie in the interval $[0, 1]$. The intersection over union (IoU) between the predicted bounding box and the ground truth box is used in confidence prediction, which reflects the likelihood that the box contains an object (objectness) and the accuracy of the boundary box. The mathematical definitions of the scoring and probability terms are as follow:

$$\text{Box confidence score } P_r(object) \cdot IoU$$
$$\text{Conditional class probability } P_r(class_i|object)$$
$$\text{Class confidence score } P_r(class_i) \cdot IoU$$
$$\text{Class confidence score} = \text{box confidence score} \times \text{conditional class probability}$$

where $P_r(object)$ is the probability that the box contains an object. IoU is the intersection over union between the predicted and ground truth boxes. $P_r(class_i)$ is the probability that the object belongs to $class_i$. $P_r(class_i|object)$ is the probability that the object belongs to $class_i$ given an object is present. The network architecture of the spatial stream contains 24 convolutional layers followed by two fully connected layers reminiscent of AlexNet and even earlier convolutional architectures. The last convolution layer is flattened, which outputs a $(7, 7, 1024)$ tensor. It performs a linear regression using two fully connected layers to make boundary box predictions and final predictions using the threshold of box confidence scores. The last loss adds the localization (the first and the second terms), confidence (the third and the fourth terms), and classification (the fifth term) losses together. The objective function is

$$\begin{aligned}\mathcal{L}_{multi} &= \lambda_{\mathbf{coord}} \sum_{i=0}^{S^2} \sum_{j=0}^{B} \mathbb{1}_{ij}^{\text{obj}} \left[(x_i - \hat{x}_i)^2 + (y_i - \hat{y}_i)^2 \right] \\ &+ \lambda_{\mathbf{coord}} \sum_{i=0}^{S^2} \sum_{j=0}^{B} \mathbb{1}_{ij}^{\text{obj}} \left[\left(\sqrt{w_i} - \sqrt{\hat{w}_i}\right)^2 + \left(\sqrt{h_i} - \sqrt{\hat{h}_i}\right)^2 \right] + \sum_{i=0}^{S^2} \sum_{j=0}^{B} \mathbb{1}_{ij}^{\text{obj}} \left(C_i - \hat{C}_i\right)^2 \\ &+ \lambda_{\text{noobj}} \sum_{i=0}^{S^2} \sum_{j=0}^{B} \mathbb{1}_{ij}^{\text{noobj}} \left(C_i - \hat{C}_i\right)^2 + \sum_{i=0}^{S^2} \mathbb{1}_{i}^{\text{obj}} \sum_{c \in \text{classes}} (p_i(c) - \hat{p}_i(c))^2, \end{aligned} \quad (2.2)$$

where $\mathbb{1}_i^{\text{obj}}$ denotes if the object appears in cell i and $\mathbb{1}_{ij}^{\text{obj}}$ denotes that the jth bounding box predictor in cell i is "responsible" for that prediction. The remaining variables are as follows: $\lambda_{\mathbf{coord}}$ and λ_{noobj} are the bounding box coordinate predictions and the confidence score predictions for boxes without objects. S is the number of cells an image is split along an axis, resulting in $S \times S$ cells. B is the number of bounding box locations predicted by each cell. A tuple of four values defines the coordinates of the bounding box (x, y, w, h) where x and y are the centers of the bounding boxes. x_i (y_i, w_i, h_i), and $\hat{x}_i$ $(\hat{y}_i, \hat{w}_i, \hat{h}_i)$ are the ground truth and prediction, respectively, of x (y, w, h). C_i is the confidence score of cell i, whereas $\hat{C}_i$ is the predicted confidence score. Finally, P_i is the conditional probability of cell i containing an object of a class, whereas $\hat{P}_i$ is the predicted conditional class probability.

2.3.2 MULTIPLE OBJECT TRACKING

MOT can generally be regarded as a multivariable estimation problem [53]. The objective of MOT can be modeled by performing MAP (maximum *a posteriori*) estimation to find the *optimal* sequential states of all the objects from the conditional distribution of the sequential states, given all the observations:

$$\widehat{\mathbf{S}}_{1:t} = \underset{\mathbf{S}_{1:t}}{\arg\max}\, P\left(\mathbf{S}_{1:t} | \mathbf{O}_{1:t}\right), \tag{2.3}$$

where $\mathbf{S}_t = (\mathbf{s}_t^1, \mathbf{s}_t^2, ..., \mathbf{s}_t^{M_t})$ denotes states of all the M_t objects in the tth frame, and $\mathbf{s}_t^i$ denotes the state of the ith object in the tth frame. $\mathbf{S}_{1:t} = \{\mathbf{S}_1, \mathbf{S}_2, \ldots, \mathbf{S}_t\}$ denotes all the sequential states of all the objects from the first frame to the tth frame. In tracking-by-detection, $\mathbf{o}_t^i$ denotes the collected observations for the ith object in the tth frame. $\mathbf{O}_t = (\mathbf{o}_t^1, \mathbf{o}_t^2, \ldots, \mathbf{o}_t^{M_t})$ denotes the collected observations for all the M_t objects in the tth frame. $\mathbf{O}_{1:t} = \{\mathbf{O}_1, \mathbf{O}_2, \ldots, \mathbf{O}_t\}$ denotes all the collected sequential observations of all the objects from the first frame to the tth frame.

Due to single-frame inputs, the spatial stream network cannot extract motion features and compute trajectories. To leverage this helpful information, we present our temporal stream network. This ConvNet model implements a tracking-by-detection MOT algorithm [49, 52] with a data association metric that combines deep appearance features. The inputs are identical to the spatial stream network using the original video. Detected object candidates (only vehicle classes) are used for tracking, state estimation, and frame-by-frame data association using SORT [49] and DeepSORT [52] – the real-time MOT method. MOT models each object's state and describes objects' motion across video frames. The tracking information allows us to stack trajectories of moving objects across several consecutive frames, which are useful cues for near-accident detection.

Estimation Model. For each target, its state is modeled as

$$\mathbf{x} = [u, v, s, r, \dot{u}, \dot{v}, \dot{s}]^T, \tag{2.4}$$

where u and v denote the 2D pixel location of the target's center. The variable s denotes the scale of the target's bounding box, and r is the aspect ratio (usually considered constant). The target's state is updated when a newly detected bounding box is associated with it by solving the velocity via a Kalman filter [50]. The target's state is predicted (without correction via the Kalman filter) if no bounding box is associated.

Data Association. To assign all new detection boxes to existing targets, we predict each target's new location (one predicted box) in the current frame and compare the IoU distance between new detection boxes and all predicted boxes (the detection-to-target overlaps), forming the assignment cost matrix. The matrix is further used in the Hungarian algorithm [51] to solve the assignment problem. One assignment is rejected if the detection-to-target overlap is less than a threshold (the minimum IoU, IoU_{min}). The IoU distances of the bounding boxes are utilized to handle the short-term occlusion caused by passing targets.

Creation and Deletion of Track Identities. When new objects enter or old objects vanish in video frames, we must add or remove certain tracks (or objects) to maintain unique identities. We treat any detection that fails to associate with existing targets as one potential untracked object. This target undergoes a probationary period to accumulate enough evidence by associating the target with detection. This strategy can prevent the tracking of false positives. The track object is removed from the tracking list if it remains undetected for T_{Lost} frames, which enables maintenance of a relatively unbounded number of trackers and reduces localization errors accumulated over a long duration.

Track Handling and State Estimation. This part is mostly identical to SORT [49]. The tracking scenario is represented as an eight-dimensional state space $(u, v, \gamma, h, \dot{x}, \dot{y}, \dot{\gamma}, \dot{h})$. The pair (u, v) represents the bounding box center location. γ is the aspect ratio, and h is the height. The rest are the bounding box's velocities relative to other bounding boxes. This representation is used to describe the constant velocity motion, $(\dot{x}, \dot{y}, \dot{\gamma}, \dot{h})$, and linear observation model, (u, v, γ, h), in Kalman filtering [50].

Data Association. To solve the frame-by-frame association problem, SORT uses the Hungarian algorithm [51], where both motion and appearance information is considered.

The (squared) Mahalanobis distance is utilized to measure the distance between newly arrived measurements and Kalman states:

$$d^{(1)}(i, j) = (\boldsymbol{d}_j - \boldsymbol{y}_i)^{\boldsymbol{T}} (\boldsymbol{S}_i)^{-1} (\boldsymbol{d}_j - \boldsymbol{y}_i), \tag{2.5}$$

where $\boldsymbol{y}_i$ and $\boldsymbol{d}_j$ denote the ith track distribution and the jth bounding box detection, respectively. The smallest cosine distance between the ith track and jth detection is considered to handle appearance information:

$$d^{(2)}(i, j) = \min \left\{ 1 - \boldsymbol{r}_j{}^T \boldsymbol{r}_k^{(i)} | \boldsymbol{r}_k^{(i)} \in \mathcal{R}_i \right\}, \tag{2.6}$$

where both motion and appearance are combined linearly, and a hyperparameter, λ, controls the influence of each:

$$c_{i,j} = \lambda\, d^{(1)}(i, j) + (1 - \lambda) d^{(2)}(i, j). \tag{2.7}$$

Matching Cascade. Previous methods tried to solve measurement-to-track associations globally, while SORT adopts a matching cascade introduced in Ref. [52] to solve a series of subproblems. In some situations, when an object is occluded for a more extended period, the subsequent Kalman filter [50] predictions would increase the uncertainty associated with the object location. Consequently, the probability mass will spread out in the state space and the observation likelihood will decrease. The measurement-to-track distance should be increased by considering the spread of probability mass. Therefore, objects seen more frequently are prioritized in the matching cascade strategy to encode the notion of probability spread in the association likelihood.

2.3.3 METRIC LEARNING FOR VEHICLE REIDENTIFICATION

Metric learning and various hand-crafted features are widely used in person reidentification problems. We apply a cosine metric learning method to train a neural network

Figure 2.5 Vehicle samples selected from the VeRi dataset [54]. All images are the same size, and we resize them to 64 × 128 for training. *Left*: Diversity of vehicle colors and types. *Right*: Variation of the viewpoints, illuminations, resolutions, and occlusions for the vehicles

for vehicle reidentification and use it to make our temporal stream more robust regarding tracking performance. Metric learning aims to solve the clustering problem by constructing an embedding in which the metric distance corresponding to the same identity is likely closer than features from different identities. The cosine metric measures the degree of similarity by calculating the cosine distance between two objects. We observe that our tracking algorithm produces switched road user identities in traffic video, especially in crowded traffic scenes or scenes with heavy occlusion. In order to generate accurate and consistent track data, we introduce a deep cosine metric learning method to learn the cosine distance between objects. The cosine distance involves appearance information that provides valuable cues for recovering identities in crowded scenes or after long-term occlusion when motion information is less discriminative. Through this deep network, the feature expression vector obtained by any object is placed in the cluster corresponding to the nearest neighbor. We trained the network on a vehicle reidentification dataset (VeRi dataset) [54] (Figure 2.5) and integrated it as the second metric measure for the assignment problem of our temporal stream.

We aim to use a unified end-to-end framework that automatically learns the best metrics. Compared to the standard distance metric (L1, L2), the learned metric can obtain more discriminative features for reidentification and is more robust to cross-view vehicle images. We adopt the architecture of Ref. [55] and train a deep network with a cosine softmax classifier that can generate feature vectors of fixed length (128) for the input images (vehicles). The cosine metric finds the nearest cluster exemplar and matches the vehicle on (sometimes far-flung) different video frames to solve tracking problems in heavy occlusion.

Given a reidentification dataset $\mathcal{D} = \{(\boldsymbol{x}_i, y_i)\}_{i=1}^{N}$ of N training images $\boldsymbol{x}_i \in \mathbb{R}^D$ and associated class labels $y_i \in \{1, \ldots, C\}$, deep metric learning is used to find a parameterized encoder function. The deep metric neural network is $\boldsymbol{r} = f_\Theta(input)$ with parameters Θ, which projects the input images $\boldsymbol{x} \in \mathbb{R}^D$ into a feature representation, $\boldsymbol{r} \in \mathbb{R}^d$, that follows a predefined notion of cosine similarity.

The cosine softmax classifier can be adapted from a standard softmax classifier and is expressed as follow:

$$p(y = k \mid \boldsymbol{r}) = \frac{\exp\left(\kappa \cdot \tilde{\boldsymbol{w}}_k^T \boldsymbol{r}\right)}{\sum_{n=1}^{C} \exp\left(\kappa \cdot \tilde{\boldsymbol{w}}_n^T \boldsymbol{r}\right)}, \tag{2.8}$$

where κ is a free-scale parameter. To generate compact clusters in the feature representation space, the first modification is to apply the ℓ_2 normalization to the final layer of the encoder network so that the representation has unit length $\|f_\Theta(\boldsymbol{x})\|_2 = 1, \forall \boldsymbol{x} \in \mathbb{R}^D$. The second modification is to normalize the weights to unit length as well, i.e., $\tilde{\boldsymbol{w}}_k = \boldsymbol{w}_k / \|\boldsymbol{w}_k\|_2, \forall k = 1, \ldots, C$. The training of the encoder network can be carried out using the cross-entropy loss, as usual. In particular, the authors in Ref. [56] have proposed accelerating the convergence of stochastic gradient descent by decoupling the length of the weight vector κ from its direction.

2.3.4 IMAGE SEGMENTATION TO DETERMINE OBJECT GAPS

To present our method fully end-to-end, we apply a learning-based dynamic distance threshold rather than a manually defined one to determine near-accident detection. A near-accident is detected if the distance between two objects in the image space is below the threshold. Using object segmentation, we introduce a threshold-learning method by estimating the gap between objects. We present a supervised learning method, using the gap distance between road users in the temporal stream to determine a near-accident while continuing to update the threshold for more convergence and precision. The straightforward way is to compute the gap using bounding boxes from detections. Still, the bounding boxes need to be more accurate in representing the boundaries of objects, particularly when rotation is involved. Therefore, we need a more compact representation and combine detections with superpixel segmentation on the video frame to get an object mask for gap distance estimation.

The main steps of our gap estimation method can be summarized as follows:

1. Extract object detections (bounding boxes) from the spatial stream network.
2. Tessellate the video frame into homogeneous superpixels.
3. Compute the IoU metric of bounding boxes and superpixels at the frame level.
4. Generate object mask per detections (using the IoU metric) and compute the gap between objects using the mask and bounding boxes.
5. Update the gap threshold while processing the video.

In our gap estimation methods, the IoU metric of boxes and superpixels is given by

$$IoU = \frac{area(B_{1:t}) \cap area(Sp_{1:t})}{area(B_{1:t}) \cup area(Sp_{1:t})}, \tag{2.9}$$

where $\mathbf{B}_t = (\mathbf{b}_t^1, \mathbf{b}_t^2, ..., \mathbf{b}_t^{M_t})$ denotes the collected detection observations for all the M_t objects in the tth frame, and $\mathbf{b}_t^i$ denotes the detection for the ith object in

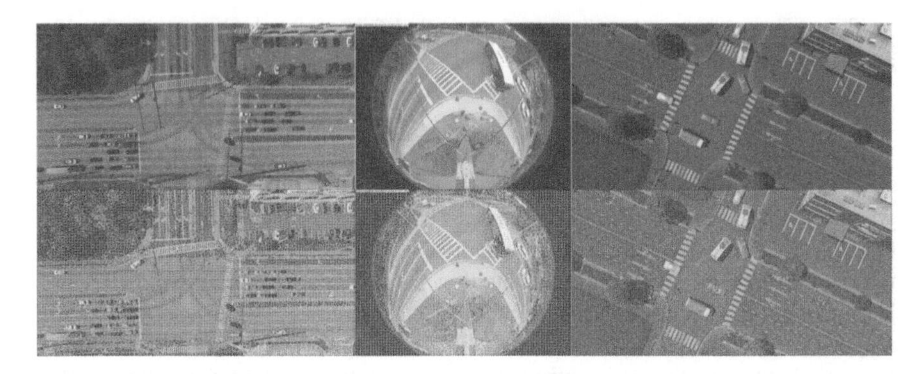

Figure 2.6 Two-dimensional image superpixel segmentation using SLIC [57]. *Top:* Original image. *Bottom-left:* SLIC segmentation for drone video (1,600 superpixels). *Bottom-middle:* SLIC segmentation for fisheye video (1,600 superpixels). *Bottom-right:* SLIC segmentation for simulation video (1,600 superpixels)

the tth frame. $\mathbf{B}_{1:t} = \{\mathbf{B}_1, \mathbf{B}_2, ..., \mathbf{B}_t\}$ denotes all the collected sequential detection observations of all the objects from the first frame to the tth frame. Similarly, $\mathbf{Sp}_t = (\mathbf{sp}_t^1, \mathbf{sp}_t^2, ..., \mathbf{sp}_t^{N_t})$ denotes all the N_t superpixels in the tth frame, and $\mathbf{sp}_t^i$ denotes the ith superpixel in the tth frame. $\mathbf{Sp}_{1:t} = \{\mathbf{Sp}_1, \mathbf{Sp}_2, ..., \mathbf{Sp}_t\}$ denotes all the collected sequential superpixels from the first frame to the tth frame.

In 2D superpixel segmentation, the popular SLIC (simple linear iterative clustering) [57] and ultrametric contour map (UCM) [58] methods have established themselves as the state of the art. Recently, we have seen deep neural networks [59] and generative adversarial networks (GANs) [60] integrated with these methods. In our architecture, we adopt the GPU-based SLIC (gSLICr) [22] approach to produce the tessellation of the image data to achieve an excellent frame rate (400 fps). SLIC is widely applicable to many computer vision applications, such as segmentation, classification, and object recognition, often achieving state-of-the-art performance. The contours forming the homogeneous superpixels are shown in Figure 2.6. In Figure 2.7, we illustrate the definition of intersection and union and present some object masks we generated from detections and superpixels.

2.3.5 NEAR-ACCIDENT DETECTION

The most typical motion cue is optical flow, which is widely utilized in video processing tasks such as video segmentation [61]. A trajectory is a data sequence containing several concatenated state vectors from tracking and an indexed sequence of positions and velocities over a given time window.

When utilizing the MOT algorithm, we compute the center of each object in several consecutive frames to form stacking trajectories as our motion representation. These stacking trajectories can provide accumulated information through image frames, including the number of objects, their motion history, and the timing of their

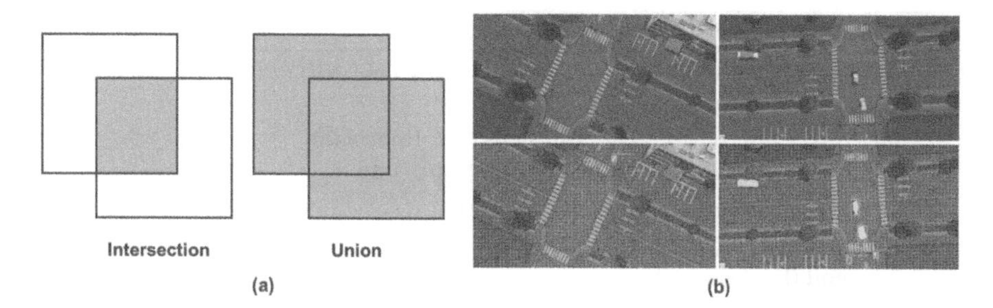

Figure 2.7 An illustration depicting the segmentation mask. (a) An illustration depicting the definitions of intersection and union. (b) *Top:* Original video frames. *Bottom:* Object detections and object masks produced by our method

interactions, such as near-accidents. We stack the trajectories of all objects for L consecutive frames, as illustrated in Figure 2.8, where $\mathbf{p}_t^i$ denotes the center position of the ith object in the tth frame. $\mathbf{P}_t = (\mathbf{p}_t^1, \mathbf{p}_t^2, \ldots, \mathbf{p}_t^{M_t})$ denotes trajectories of all the M_t objects in the tth frame. $\mathbf{P}_{1:t} = \{\mathbf{P}_1, \mathbf{P}_2, \ldots, \mathbf{P}_t\}$ denotes the sequence of trajectories of all the objects from the first frame to the tth frame. For L consecutive frames, the stacking trajectories are defined by the following sequence:

$$\mathbf{P}_{1:L} = \{\mathbf{P}_1, \mathbf{P}_2, ..., \mathbf{P}_L\}, \quad \mathbf{P}_{L+1:2L} = \{\mathbf{P}_{L+1}, \mathbf{P}_{L+2}, \ldots, \mathbf{P}_{2L}\}, \quad \ldots \tag{2.10}$$

$\mathbf{O}_t = (\mathbf{o}_t^1, \mathbf{o}_t^2, \ldots, \mathbf{o}_t^{M_t})$ denotes the collected observations for all the M_t objects in the tth frame. $\mathbf{O}_{1:t} = \{\mathbf{O}_1, \mathbf{O}_2, \ldots, \mathbf{O}_t\}$ denotes all the collected sequential observations of all the objects from the first frame to the tth frame. We use a simple detection algorithm that finds collisions between simplified forms of the objects using the center of bounding boxes.

Our algorithm is depicted in Algorithm 1. Once a collision is detected, we set the region covering collision-associated objects as a new bounding box with a class

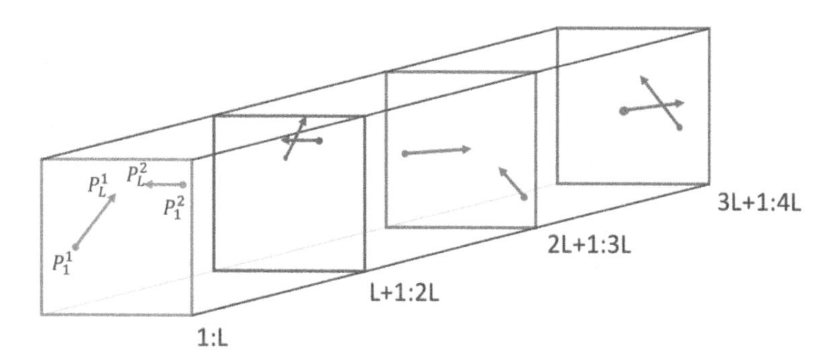

Figure 2.8 The stacking trajectories extracted from multiple object tracking of the temporal stream. Consecutive frames and the corresponding displacement vectors are shown with the same color

Algorithm 1: Collision Detection Algorithm

```
Input: current frame t_current, collision state list Collision
Output: collision state list Collision
for t_L ← t_previous to t_current in steps of L frames do
    for each pair of object trajectory (p^1_{:t_L}, p^2_{:t_L}) do
        if (p^1_{:t_L} intersects p^2_{:t_L} as of t_L) then
            add o_1, o_2 to Collision
        end if
    end for
    if (Collisions) then
        t_previous ← t_L; return TRUE
    end if
end for
t_previous ← t_d; return FALSE
```

probability of near-accident as 1. We can obtain a final confidence measure of near-accident detection by averaging the near-accident probabilities from the spatial stream network and the temporal stream network.

2.4 EXPERIMENTS

Here, we present qualitative and quantitative evaluations regarding object detection, MOT, and near-accident detection performance. We also compare other methods with our framework.

2.4.1 A TRAFFIC NEAR-ACCIDENT DATASET (TNAD)

To our best knowledge, we know of no comprehensive traffic near-accident dataset containing top-down-view videos such as drone or UAV videos or omnidirectional camera videos for traffic analysis. Therefore, we have built our traffic near-accident dataset (TNAD), depicted in Figure 2.9. Intersections tend to experience more near-accidents and more potentially severe ones due to factors such as angles and turning collisions. The traffic near-accident dataset contains three types of video data from traffic intersections that could be utilized for near-accident detection and other traffic surveillance tasks, including turn movement counting.

The first type is drone video that monitors an intersection with a top-down view. The second type of intersection video is real traffic video acquired by omnidirectional fisheye cameras that monitor small or large intersections. They are widely used in transportation surveillance. These video data can be directly used as input for our vision-based intelligent framework. Furthermore, preprocessing and fisheye correction can be applied for better surveillance performance. The third type of video is video game-engine simulations that train on near-accident samples as they are accumulated. The dataset consists of 106 videos with a total duration of over

Figure 2.9 Samples of traffic near-accident: Our data consists of many diverse intersection surveillance videos and near-accidents (cars and motorbikes). Yellow rectangles and lines represent the same object in video from multiple cameras. White circles represent the near-accident regions

75 minutes, with frame rates between 20 and 50 fps. The drone and fisheye surveillance videos were recorded in Gainesville, FL at several intersections. Our videos are more challenging than videos in other datasets for the following reasons:

- *Diverse intersection scene and camera perspectives:* The intersections in drone video, fisheye surveillance video, and simulation video are very diverse. Additionally, the fisheye surveillance video has distortion, and a fusion technique is needed for multi-camera fisheye videos.
- *Crowded intersection and small object:* The number of moving cars and motorbikes per frame is large, and these objects are smaller than in normal traffic video.
- *Diverse accidents:* Accidents involving cars and motorbikes are all included.
- *Diverse lighting conditions:* Lighting conditions such as daylight and sunset are included.

We manually annotated the spatial and temporal locations of near-accidents, the still or moving objects, and their vehicle class in each video. Thirty-two videos with sparsely sampled frames (only 20% of the frames in these 32 videos are used for supervision) were used, but only for training the object detector. The remaining 74 videos were used for testing.

2.4.2 FISHEYE AND MULTI-CAMERA VIDEO

We have large amounts of fisheye traffic videos from across the city. The fisheye surveillance videos were recorded from real traffic data in Gainesville. We collected 29 single-camera fisheye surveillance videos and 19 multi-camera fisheye surveillance videos monitoring a large intersection. We conducted two experiments, one directly using these raw videos as input for our system and another in which we first preprocessed the video to correct fisheye distortion and then fed them into our system. As the original surveillance video has many visual distortions, especially near the circular boundaries of the cameras, our system performed better on these after preprocessing. In this chapter, we do not discuss issues related to fisheye unwarping, leaving these for future work.

For large intersections, two fisheye cameras placed to opposite each other are used for surveillance, showing almost half the roads and real traffic. We apply a simple *object-level* stitching method by assigning the object identifier for the same objects across the left and right videos using similar features and appearing and vanishing positions.

2.4.3 MODEL TRAINING

We adopt Darknet-19 [52] as the backbone network for our classification and detection tasks. To achieve robust and precise tracking of objects, we integrate the DeepSORT, algorithm into our framework. DeepSORT leverages a data association metric that effectively combines deep appearance features, allowing it to link detections across frames and maintain accurate object trajectories, even in challenging scenarios such as occlusions and crowded scenes.

We implement our framework on Tensorflow and perform multiscale training and testing with a single GPU (NVIDIA Titan X Pascal). Training a single spatial convolutional network takes one day on our system with one NVIDIA Titan X Pascal card. We use the same training strategy for classification and detection training as YOLO9000 [52]. We train the network on our dataset with four classes of vehicle (motorbike, bus, car, and truck) for 160 epochs, using stochastic gradient descent with a starting learning rate of 0.1 for classification and 10^{-3} for detection (dividing it by 10 at 60 and 90 epochs.), weight decay of 0.0005, and momentum of 0.9 using the darknet neural network framework [52].

2.4.4 QUALITATIVE RESULTS

We present some example experimental results of object detection, MOT, and near-accident detection on our traffic near-accident dataset (TNAD) for drones, fisheye, and simulation videos. For object detection (Figure 2.10), we present some detection results of our spatial network with multiscale training based on YOLOv2 [47]. These visual results demonstrate that the spatial stream is able to detect and classify road users with reasonable accuracy, even if the original fisheye videos have significant distortions and occlusions. Unlike drone and simulation videos, we have added pedestrian detection on real traffic fisheye videos. From Figure 2.10, our detector performs well on pedestrian detection for daylight and dawn fisheye videos. The vehicle detection capabilities are good on top-down view surveillance videos, even for small objects. In addition, we can achieve a fast detection rate of 20–30 frames per second. Overall, this demonstrates the effectiveness of our spatial neural network.

For MOT (Figure 2.11), we present a comparison of our temporal network based on DeepSORT [52] with Urban Tracker [62] and TrafficIntelligence [63]. We also present more visual MOT comparisons of our baseline temporal stream (DeepSORT)

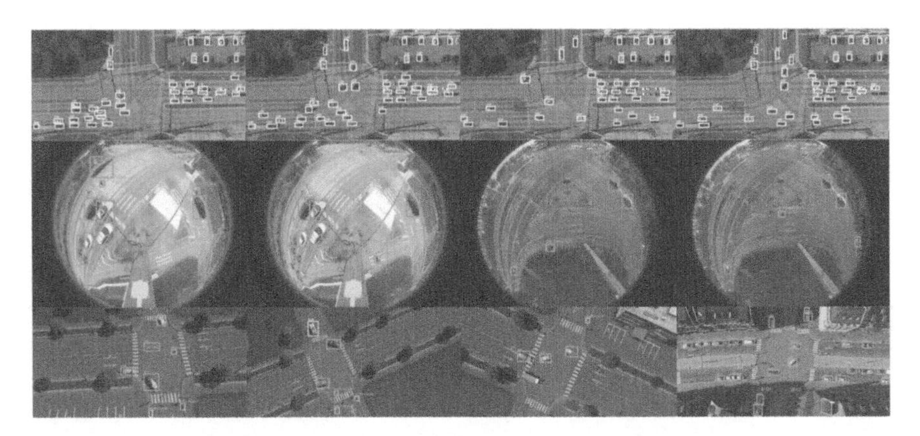

Figure 2.10 Object detection result samples of our spatial network on TNAD dataset. *Top:* Samples for drone video. *Middle:* Samples for fisheye video. *Bottom:* Samples for simulation video

with cosine metric learning-based temporal stream (Figure 2.12). For the tracking part, we use a tracking-by-detection paradigm; thus, our methods can handle still objects and measure their state. This is especially useful because Urban Tracker [62] and TrafficIntelligence [63] can only track moving objects. On the other hand, Urban Tracker [62] and TrafficIntelligence [63] can compute dense trajectories of moving objects with reasonable accuracy, but they have slower tracking speed – around 1 frame per second. Our two-stream convolutional networks can do spatial and temporal localization for accident detection for diverse accident regions involving cars and motorbikes. The three subtasks (object detection, MOT, and near-accident detection) can consistently achieve real-time performance at a high frame rate – 40

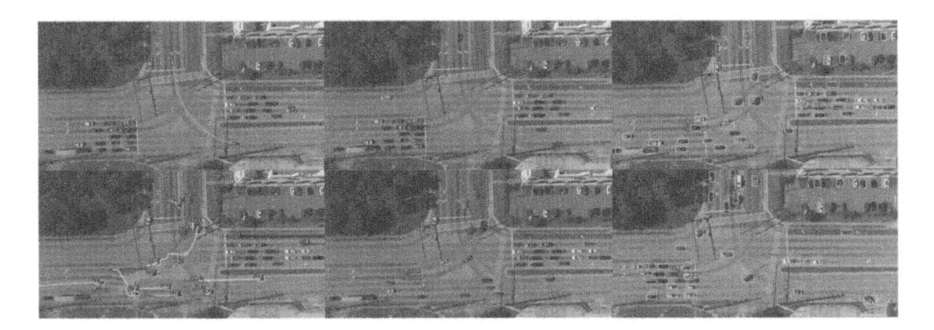

Figure 2.11 Tracking and trajectory comparison with Urban Tracker [62] and TrafficIntelligence [63] on drone videos of TNAD dataset. *Left:* Tracking results of Urban Tracker [62] (BSG with Multilayer and Lobster Model. *Middle:* Tracking results of TrafficIntelligence [63]. *Right:* Tracking results of our temporal stream network

Figure 2.12 Multiple object tracking (MOT) comparisons of our baseline temporal stream (SORT) with cosine metric learning-based temporal stream. *Row 1:* Results from SORT (ID switching issue). *Row 2:* Results from cosine metric learning + SORT (ID consistency). *Row 3:* Results from SORT (ID switching issue). *Row 4:* Results from cosine metric learning + SORT (ID consistency)

to 50 frames per second – and this depends on the frame resolution (e.g., 50 fps for 960 × 480 image frames).

In Figure 2.12, we demonstrate the effectiveness of applying cosine metric learning with the temporal stream for solving ID switching (and for the case where new object IDs emerge). For example, the first row shows that the ID of the green car (inside the red circle) got changed from 32 to 41. The third row shows that the IDs of the red car and a pedestrian (inside the red circle) were changed (from 30 to 43 and 38 to 44, respectively). The second and fourth rows demonstrate that with cosine metric learning, the IDs of tracks are kept consistent. In Figure 2.13, we show an example of a more accurate and compact road user mask generated by superpixel segmentation and detections for the learning-based gap estimation. With a segmentation mask, we can estimate the gap distance between road users more accurately than directly using object detections. Segmentation would be more beneficial for fisheye videos where the distortion is significant and occlusion is heavier. For near-accident detection (Figure 2.14), we present final near-accident detection results along with tracking and trajectories using our two-stream convolutional networks method. Overall, the qualitative results demonstrate the effectiveness of our spatial and temporal networks.

Figure 2.13 Object segmentation using detections and superpixels. *Top-left:* Original image. *Top-middle:* Superpixels generated by SLIC [57]. *Top-right:* Final object mask. *Bottom-left:* Object detections. *Bottom-middle:* Colored superpixels generated by SLIC [57]. *Bottom-right:* Final binary object mask

Table 2.1
Quantitative Evaluation of Timing Results for the Tested Methods

Methods	GPU	Drone	Simulation		Fisheye
		1,920 × 960	2,560 × 1,280	3360 × 1,680	1,280 × 960
SLIC segmentation	Nvidia Titan V	300 fps	178 fps	110 fps	400 fps
Two-stream CNNs	Nvidia Titian V	43 fps	38 fps	33 fps	50 fps

2.4.5 QUANTITATIVE RESULTS

2.4.5.1 Speed Performance

We present timing results for the tested methods in Table 2.1. All experiments have been performed on a single GPU (NVIDIA Titan V). The GPU-based SLIC segmentation [57] has excellent speed and runs from 110 to 400 fps on videos of different resolutions. The two-stream CNNs (object detection, MOT, and near-accident detection) can achieve real-time performance at a high frame rate, 33–50 fps, depending on the frame resolution.

2.4.5.2 Cosine Metric Learning

We present some quantitative results for gap estimation and cosine metric learning in Figure 2.15. The results for cosine metric learning have been established after training the network for a fixed number of steps (100,000 iterations). The batch size was set to 120 images, and the learning rate was 0.001. All configurations after training have fully converged, as depicted in Figure 2.15.

Triplet Loss. The triplet loss for cosine metric learning network [64] is defined by

$$\mathcal{L}_t(\boldsymbol{r}_a, \boldsymbol{r}_p, \boldsymbol{r}_n) = \left\{\|\boldsymbol{r}_a - \boldsymbol{r}_n\|_2 - \|\boldsymbol{r}_a - \boldsymbol{r}_p\|_2 + m\right\}_+, \tag{2.11}$$

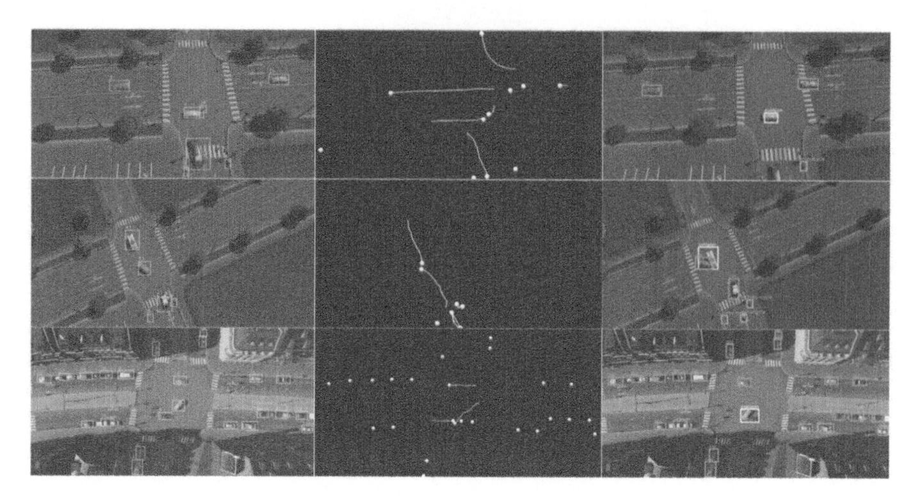

Figure 2.14 Sample results of tracking, trajectory, and near-accident detection of our two-stream convolutional networks on simulation videos of TNAD dataset. *Left:* Tracking results from the temporal stream. *Middle:* Trajectory results from the temporal stream. *Right:* Final near-accident detection results from the two-stream convolutional networks

where $\boldsymbol{r}_a$, $\boldsymbol{r}_p$, and $\boldsymbol{r}_n$ are three samples in which a positive pair, $y_a = y_p$, and a negative pair, $y_a \neq y_n$, are included. With a predefined margin $m \in \mathbb{R}$, the triplet loss demands the distance between the positive and negative is larger than it. In this experiment, we introduce a soft margin–based triplet loss where we replace the hinge of the original triplet loss [65] by a soft plus function, $\{x + m\}_+ = \log(1 + \exp(x))$, to resolve nonsmoothness [66] issues. Further, to avoid potential issues in the sampling strategy, we directly generate the triplets on GPU as proposed by Ref. [65].

Magnet Loss. Magnet loss is defined as a likelihood ratio measure that demands the separation of all samples away from the means of other classes. Instead of using a multimodal form in its original proposition [66], we use a unimodal, adapted version of this loss to fit the vehicle reidentification problem:

$$\mathcal{L}_{\mathrm{m}}(y, \boldsymbol{r}) = \left\{ -\log \frac{\mathrm{e}^{-\frac{1}{2\hat{\sigma}^2}\|\boldsymbol{r}-\hat{\boldsymbol{\mu}}_y\|_2^2 - m}}{\sum_{k\in\bar{C}(y)} \mathrm{e}^{-\frac{1}{2\hat{\sigma}^2}\|\boldsymbol{r}-\hat{\boldsymbol{\mu}}_k\|_2^2}} \right\}_+, \tag{2.12}$$

where $\bar{C}(y) = \{1, \ldots, C\} \setminus \{y\}$, and m is the predefined margin parameter, $\hat{\boldsymbol{\mu}}_y$ is used to represent the sample mean of class y, and $\hat{\sigma}^{2}$ denotes the variance of each sample separated away from their class mean. They are all established for each batch individually on GPU.

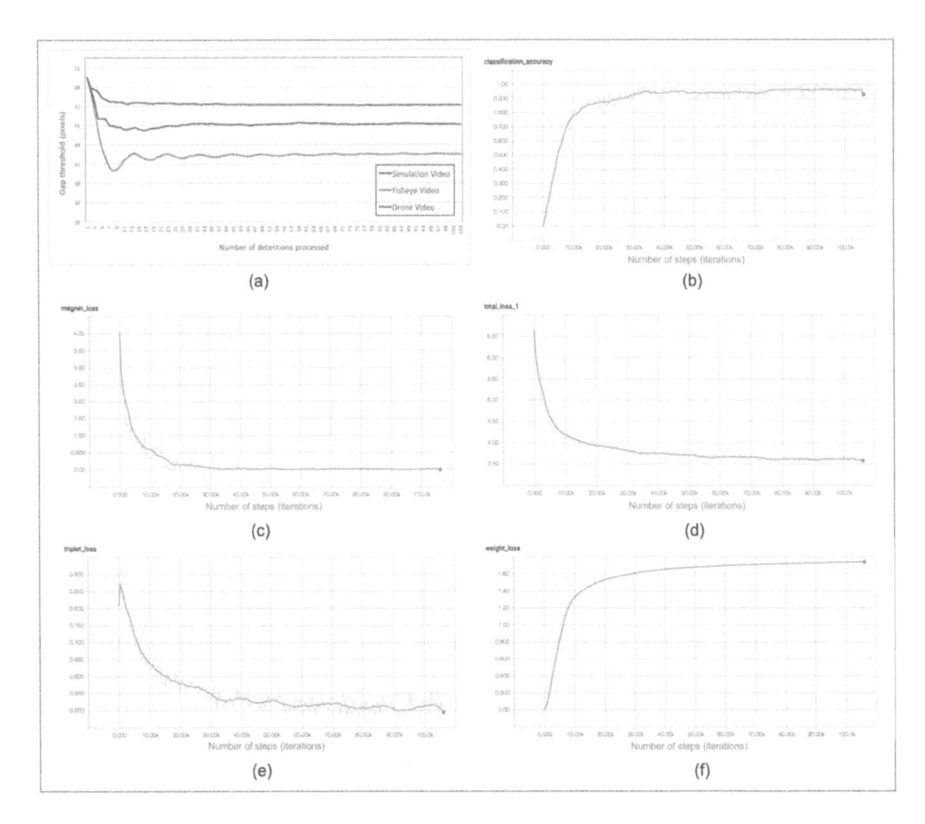

Figure 2.15 Gap estimation and cosine metric learning. (a) Evolution of gap threshold for three types of videos. (b) Classification accuracy for training VeRi dataset. (c) Evolution of total loss for training. (d) Evolution of magnet loss for training. (e) Evolution of triplet loss for training. (f) Evolution of weight loss for training

2.4.5.3 Object Segmentation

For learning the gap threshold with superpixel segmentation and detections, the plot (Figure 2.15a) demonstrates the converging evolution of gap distance for three types of video. The initial gap threshold is set to be 50 pixels; the average gap thresholds from experiments are about 47 pixels, 45 pixels, and 42 pixels for simulation video, fisheye video, and drone video, respectively.

The evaluation method we use, for instance, segmentation, is quite similar to object detection, except that we now calculate the IoU of masks instead of bounding boxes. **Mask Precision, Recall, and F1 Score.** To evaluate our collection of predicted masks, we'll compare each of our predicted masks with each of the available target masks for a given input.

- A true positive is observed when a prediction-target mask pair has an IoU score that exceeds some predefined threshold (we set it to 0.7).
- A false positive indicates a predicted object mask had no associated ground truth object mask.

Table 2.2
Quantitative Evaluation of All Tasks for Our Two-stream Method

Task or Methods	Data	Precision	Recall	F1 Score
Object detection	Simulation	0.92670	0.95255	0.93945
	Fisheye	0.93871	0.87978	0.90829
Multiple object tracking	Simulation	0.89788	0.86617	0.88174
	Fisheye	0.91239	0.84900	0.87956
Multiple object tracking (cosine metric)	Simulation	0.91913	0.90310	0.91105
	Fisheye	0.92663	0.87725	0.90127
SLIC segmentation mask	Simulation	0.93089	0.81321	0.86808
Near-accident detection (spatial stream)	Simulation	0.90395	0.86022	0.88154
Near-accident detection (temporal stream)	Simulation	0.83495	0.92473	0.84108
Near-accident detection (two-stream)	Simulation	0.92105	0.94086	0.93085

- A false negative indicates a ground truth object mask had no associated predicted object mask.

Since our framework has three tasks and our dataset is quite different from other object detection datasets, tracking datasets, and near-accident datasets such as dashcam accident dataset [67], it isn't easy to compare the individual quantitative performance for all three tasks with other methods. One of our motivations was to propose a vision-based solution for ITS; therefore, we focus more on near-accident detection and present a quantitative analysis of our two-stream convolutional networks. In Table 2.2, we present quantitative evaluations of three subtasks: object detection; MOT (with and without cosine metric learning); SLIC segmentation (mask vs. bounding box); and near-accident detection (spatial stream only, temporal stream only, and two-stream model). We present evaluations regarding object detection and MOT for the fisheye video. The simulation videos are for training and testing with more near-accident samples, and we have 57 simulation videos totaling over 51,123 video frames. We sparsely sample only 1,087 frames from them for training processing. We present the analysis of near-accident detection for 30 testing videos (18 had positive near-accident; 12 had negative near-accident).

Near-accident Detection Precision, Recall, and F1 Score. We'll compare our predicted detections with each ground truth for a given input to evaluate our prediction of near-accidents.

- A true positive is observed when a prediction–target detection pair has an IoU score that exceeds some predefined threshold (we set it to 0.7).
- A false positive indicates a predicted near-accident had no associated ground truth near-accident.
- A false negative indicates a ground truth near-accident had no associated predicted detections.

2.5 DISCUSSION

We have proposed a two-stream convolutional network architecture that performs real-time detection, tracking, and near-accident detection for vehicles in traffic video data. The two-stream convolutional network comprises a spatial and temporal stream network. The spatial stream network detects individual vehicles and likely near-accident regions at the single-frame level by capturing appearance features with a state-of-the-art object detection method. The temporal stream network leverages motion features of detected candidates to perform MOT and generates individual trajectories of each tracked target. We detect near-accidents by incorporating appearance and motion features to compute probabilities of near-accident candidate regions. Experiments have demonstrated the advantage of our framework with an overall competitive qualitative and quantitative performance at high frame rates. Future work will include image stitching methods deployed on multi-camera fisheye videos.

3 Near-miss Detection

3.1 INTRODUCTION

We develop a novel workflow for analyzing vehicular and pedestrian traffic at an intersection, beginning with ingesting video data and controller logs, followed by data storage and processing to generate dominant and anomalous behavior at the intersection, which helps in various applications such as near-miss detection [68], which is explained in detail in this chapter. The critical contributions of the work in this chapter are as follows:

1. We develop a new distance measure for computing distances between trajectories. We develop an offline, two-level hierarchical clustering scheme using this distance measure. At the first level, the trajectories are clustered based on their direction of movement. At the second level, spectral clustering is applied. Clustering helps us to detect the outliers automatically.
2. We show how new insights and perspectives into the trajectory data are possible by joining the trajectory database with the signal data. For example, the combined data can detect signal violations, count the number of vehicles entering the intersection on a yellow light, and several other useful behavior patterns.

Extensive results are provided on video and signal data collected at an intersection. These results demonstrate that video and signal timing information is useful in quantifying the following:

1. Safety of pedestrians and bicyclists by studying the nature of the anomalous vehicle trajectories and also the statistics of occurrence of these anomalies (counts of anomalies depending upon the hour and day of the week)
2. Effective tuning of signal timing based on demand profiles. It also helps us compare the new technologies, such as video-based monitoring, and existing technologies, such as induction loops.

The rest of the chapter is organized as follows. Section 3.2 describes the trajectory generation in brief, and Section 3.3 describes the recording of the current signal state of an intersection. Finally, Section 3.4 presents a detailed comparison of the candidate distance measures. Sections 3.7–3.9 presents an exhaustive treatise on near-miss detection.

3.2 TRAJECTORY GENERATION

Video processing software processes object locations frame by frame from a video and outputs the location coordinates along with the corresponding timestamp. The video is captured by a camera installed at an intersection. To accurately locate the

DOI: 10.1201/9781003431176-3

coordinates of an object, the video processing software must account for the different types of distortions that creep into the system. For example, a fisheye lens would have significant radial distortion. After taking into account the intrinsic and extrinsic properties of the camera, a mapping is created, which is used by the video processing software to map observed coordinates to modified coordinates that are nearly free of any distortion. To represent the location of a 3D object using a dimensionless point, one looks to find the object's center of mass. A bounding box is drawn, enclosing the object. The center of the box is approximated to be the object's center of mass. After generating timestamped trajectory coordinates, the software computes the modified coordinates in a distortion-free space. Finally, it uses these modified coordinates to calculate other properties, such as speed and direction of movement.

3.3 SIGNALING STATUS

An intersection almost always has traffic lights to control traffic flow safely. The signal changes from green to yellow to red are events captured in controller logs called signal data. To specify a particular signal and, in a more general sense, the direction of movement, the traffic engineers define a standard that assigns phases 2 and 6 to the two opposite directions of the major street and 4 and 8 to those of the minor street. Figure 3.1 shows these phase numbers and the phase numbers for the turning vehicles and pedestrians.

Our application stores the current signal phase in a compact six-digit hexadecimal encoding. To explain the formatting, let us consider the corresponding 24-bit binary equivalent. The bits 1–8 are programmed to be 1 if the corresponding phase is green and 0 otherwise. Similarly, bits 9–16 and 17–24 are reserved for programming the yellow and red status for the eight-vehicle phases, respectively. For example, green on phases 2 and 6 at an intersection would have a binary encoding of *0100 0100* for the first 8 bits, the next eight bits would be *0000 0000* for yellow, and the last set of 8 bits for red would be one where the second and sixth bits are 0 represented as *1011 1011*. Thus, the overall 24-bit binary representation of the current signaling state is *0100 0100 0000 0000 1011 1011*, or *4400bb* (in hexadecimal format).

3.4 COMPARING TRAJECTORIES

The first step toward clustering a set of trajectories is applying a suitable distance measure that will, for any two trajectories, tell how close the trajectories are to each other in space and time. Two potential candidates for distance measures of the intersection trajectories are Euclidean distance (ED) and dynamic time warping (DTW). Among these, ED is the square root of the sum of the squared length of vertical or horizontal hatched lines. The disadvantage of using ED is that it cannot calculate their distance reliably for trajectories of different sizes.

DTW can compute distances between trajectories when they vary in time, speed, or path length. Although DTW utilizes a dynamic programming approach for an optimal distance and has time complexity $O(N^2)$, where N is the number of coordinates in the two trajectories, there are approximate approaches such as FastDTW that realize a near-optimal solution and has a space and time complexity of $O(N)$. FastDTW is

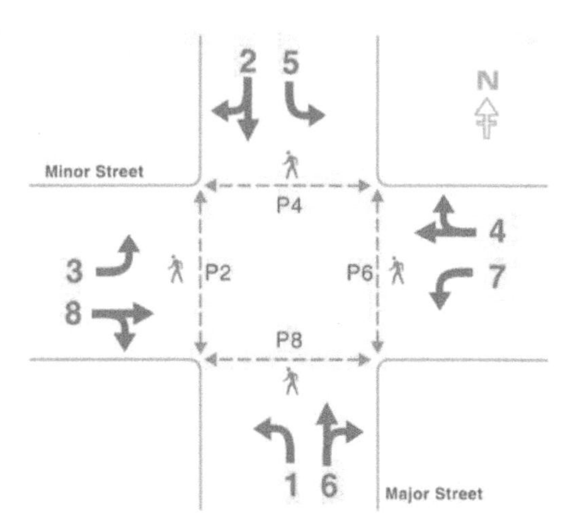

Figure 3.1 Phase diagram showing vehicular and pedestrian movement at four-way intersections. The solid gray arrows show vehicle movements while the blue dotted arrows show pedestrian movements [69]

based on a multilevel iterative approach. FastDTW returns a distance and a list of pairs of points, also known as the warp path. A pair of points consist of coordinates on the first and second trajectories and represent the best match between the points after the trajectories are warped. The distance returned by FastDTW is the sum of the distances between each pair of points on the path. Quite naturally, the distance is negligible if the trajectories occur in the same geographical coordinates and are traversed at similar speeds.

DTW and FastDTW work well for trajectories entirely captured by the sensor system. In reality, the sensor system and the processing software may not wholly capture the trajectory. In that case, the distance the dynamic time-warping algorithm returns does not represent the actual distance. Figure 3.2 illustrates the alignment obtained between coordinates of a pair of trajectories using the DTW algorithm. Figure 3.3 highlights two example trajectories for which DTW returns a high value for the distance suggesting the tracks are dissimilar, which they are not. For example, for the two tracks going straight, if the starting portion of one of the tracks is truncated due to a processing error, as shown in Figure 3.4, the distance between the tracks will be $\sqrt{\overline{AF}^2 + \overline{BF}^2 + \overline{CF}^2 + \overline{DG}^2 + \overline{EH}^2}$. Thus, the distance computed results in a high value, which often falls in the distance range between two unrelated trajectories. Hence, we developed a new distance measure by utilizing the warp path returned by the FastDTW.

Figure 3.2 Alignment obtained for two tracks using the dynamic time warping algorithm

3.5 CLUSTERING

We describe our trajectory clustering method in this section. There are two main components to clustering – the first is to use an efficient distance measure for the trajectories to be clustered, and the second is to use a clustering algorithm that uses the

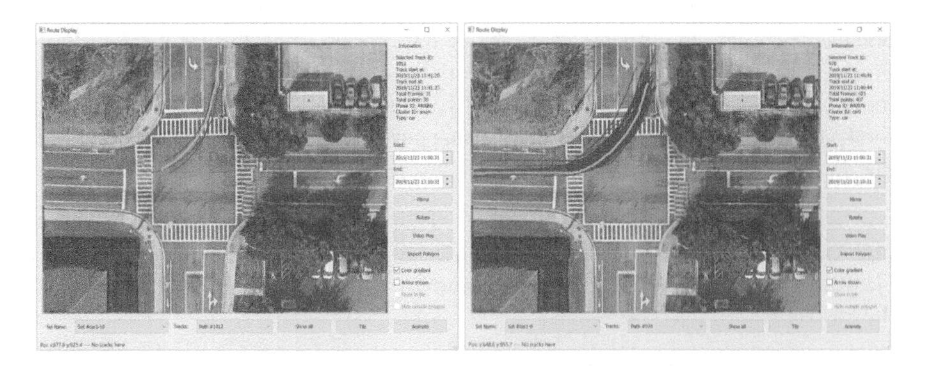

Figure 3.3 An example where two similar trajectories have a high distance value when DTW/FastDTW is directly used to compute distance. This is potentially due to differential tracking of vehicles due to occlusion

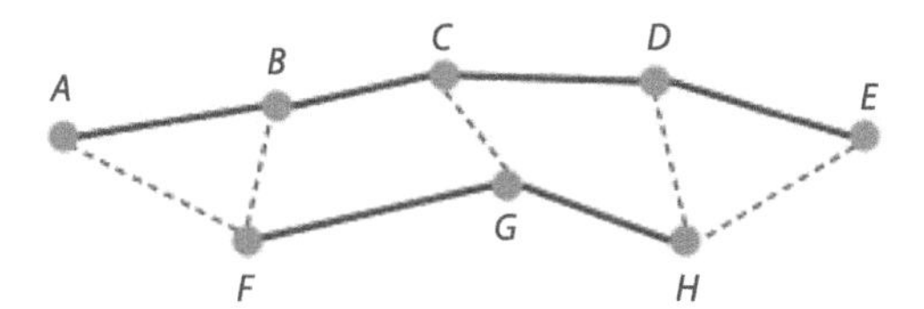

Figure 3.4 Two trajectories represented by *ABCDE* and *FGH*. The dashed lines show the point correspondence (warp path) obtained using FastDTW. The trajectory *FGH* is shorter because the beginning part of the trajectory was not captured

distance measure to create clusters of trajectories that behave similarly. We describe the novel distance measure developed as part of this work in Section 3.5.1 and present our clustering algorithm in Section 3.5.2.

3.5.1 DISTANCE MEASURE

The new distance measure developed in this section applies to trajectories captured using real-time video processing. The first step in the computation of the distance measure is to obtain the warp path using a time-warping algorithm, e.g., FastDTW. Triangles are constructed using the warp path as shown in Figure 3.4, where the triangles are $\triangle ABF$, $\triangle BCF$, $\triangle CDF$, $\triangle DFG$, $\triangle DEG$, and $\triangle EGH$. Since the coordinates of the vertices (A, B, C, D, E, F, G, and H) are known, the area of each may be computed using the following formula from coordinate geometry:

$$Area = \left| \frac{a_x(b_y - c_y) + b_x(c_y - a_y) + c_x(a_y - b_y)}{2} \right|, \tag{3.1}$$

where the vertices of the triangle have coordinates: (a_x, a_y), (b_x, b_y), and (c_x, c_y). The sum of the area of all triangles is computed, and, finally, the distance D_{ij} between two trajectories T_i and T_j is computed as folows:

$$D_{ij} = (\sum_{k=1}^{n} Area_k)/\overline{L_{ij}}, \tag{3.2}$$

where $Area_k$ is the area of the kth triangle and $\overline{L_{ij}}$ is the average length of the two trajectories, T_i and T_j, and n is the total number of triangles. If there is no warping, and the number of matched pairs is m, then $n = 2m$, by construction. However, if there is warping, then $n < 2m$. D_{ij} is the average perpendicular distance between the trajectories, which intuitively is the average height of the triangles.

To get a more accurate local distance measure, we segment the trajectories and compute the distance of the starting and the finishing segments. Let SD_{ij} and FD_{ij} be the distances of the starting and the finishing segments, respectively. Then, SD_{ij} is the distance between the first pair of matching points that are not warped (CF and DG in Figure 3.4), and, correspondingly, FD_{ij} is the distance between the last pair of matching points that are not warped (DG and EF in Figure 3.4). SD_{ij} may be computed as the average height of triangles $\triangle CFG$ and $\triangle CDG$, and FD_{ij} as that of triangles $\triangle DGH$ and $\triangle DEH$. Thus, we use triplets of $(D_{ij}, SD_{ij}, FD_{ij})$ to represent the distance between two trajectories. If required, these three portions of the distance measure can be suitably weighted for computing a scalar distance.

3.5.1.1 Similarity Matrix

The similarity matrix, S, is computed from the distance measures by setting up empirical thresholds for the magnitude of the distance between two trajectories. For example, given two trajectories T_i and T_j, if their average distance D_{ij} from Equation 3.2 is less than a threshold ϵ_x and the start section distance SD_{ij} is less than a threshold ϵ_y and the last section distance FD_{ij} is less than a threshold ϵ_z, then T_i and T_j are considered similar. In that case, the corresponding entry in the similarity matrix would be $S_{ij} = S_{ji} = 1$. If any distance value exceeds the corresponding threshold ϵ_x, ϵ_y, or ϵ_z, the trajectories would be considered dissimilar, and in that case, $S_{ij} = S_{ji} = 0$. In this manner, the similarity matrix is computed and is ready to be used in spectral clustering.

3.5.2 HIERARCHICAL CLUSTERING

Clustering a large set of N trajectories is a $O(N^2)$ operation because pairwise distance needs to be computed to prepare a distance matrix. A two-level hierarchical clustering scheme is proposed here to address the quadratic complexity. Clustering at the first level partitions the trajectories into homogeneous clusters based on their direction of movement. Then spectral clustering is applied to cluster the trajectories for each movement direction separately and to detect anomalies. The clustering scheme is explained in detail in this section.

3.5.2.1 Partitioning Trajectories Based on Movement Phase

Any object at an intersection must obey the traffic rules; hence, its trajectory is constrained in time and space. One of the goals of the software is to detect traffic violations and the underlying causes to make the intersection safer ultimately. These violations may appear as spatial outliers or as timing violations. Figure 3.1 shows the phases for pedestrian and vehicular movements. A given set of trajectories is partitioned into eight bins aligned with the eight phases and into four additional bins corresponding to the right turns for phases 2, 4, 6, and 8.

Because the trajectories are essentially a series of coordinates, it is possible to use basic vector algebra and trigonometry to get their general direction. For example, the spatial coordinates of a track T_i is given as $(x_1, y_1), (x_2, y_2), \ldots, (x_n, y_n)$. Let $\boldsymbol{A} = [x_1, y_1]$ and $\boldsymbol{B} = [x_n, y_n]$ be two vectors connecting the origin to the start and end points, respectively, of T_i, with their direction away from the origin. Let $\boldsymbol{AB}$ be a vector connecting the start and the end points, with its head at the end point, (x_n, y_n). Then, using the rules of vector addition, $\boldsymbol{A} + \boldsymbol{AB} = \boldsymbol{B}$, which implies $\boldsymbol{AB} = \boldsymbol{B} - \boldsymbol{A}$. Thus, $\boldsymbol{AB} = [x_n - x_1, y_n - y_1]$.

```
phase2 WB
SN (562, 880), (560, 149)
NS (560, 149), (562, 880)
WE (84, 536), (962, 547)
EW (962, 547), (84, 536)
phase2stopbar (815, 153), (819, 707)
phase4stopbar (244, 739), (825, 750)
phase6stopbar (330, 155), (287, 726)
phase8stopbar (287, 240), (804, 231)
```

Figure 3.5 Snippet of a configuration file that shows the user inputs needed for an intersection. The coordinates may be obtained using a visualization tool [70]

Once constructed, a trajectory vector may be compared with a reference vector to obtain the direction of the trajectory. The start and end points of these reference vectors may be obtained using CAD tools supported by visualization software, and the coordinates of the start and end points are specified in a configuration file.

The cosine of the acute angle between a trajectory vector $\boldsymbol{u}$ and a reference vector $\boldsymbol{v}$ is given by

$$\cos\theta = \frac{\boldsymbol{u} \cdot \boldsymbol{v}}{\|\boldsymbol{u}\| \, \|\boldsymbol{v}\|}. \tag{3.3}$$

When the directions nearly match, the value of the cosine is close to 1.0.

A snippet of a configuration file used to specify the reference vectors is shown in Figure 3.5. The user can specify the direction of the phase 2 movement using the phase 2 parameter of this file, which has a value WB (westbound) in this example. The other phases, as described in Figure 3.1, may be derived with reference to the phase 2 direction. The parameters SN, NS, WE, and EW specify the start and end points of reference vectors for through lanes along south–north, north–south, west–east, and east–west directions, respectively. Figure 3.6 shows the corresponding reference vectors. The start and end point coordinates of the stop bars are also needed to differentiate left- and right-turn movements that otherwise align with each other. An example of this is given later in this section.

Given a trajectory, the cosine value in Equation 3.3 may be computed for the trajectory vector and the reference vectors, and if the value calculated is close to 1 for any of these vectors, the trajectory may be assigned to the corresponding phase (one of 2, 4, 6, or 8). For any intersection, the coordinates of the reference vectors have to be determined once, and the configuration file may be reused over time until the geometry of the intersection is changed.

3.5.2.2 Clustering Trajectories in a Partition

The next step in the hierarchical clustering scheme is to cluster the trajectories with the same movement direction. The spectral clustering algorithm is applied in this step. Prior experimentation with a simple K-means clustering approach for this problem highlights the benefits of using spectral clustering instead. A pure K-means

Figure 3.6 Pictorial representation of information in the configuration file

algorithm requires user input for the number of possible clusters, which is impossible for the user to know in advance. Spectral clustering, through its smart use of standard linear algebra methods, gives the user objective feedback about possible clusters and further accentuates the trajectories' features to make cluster separation easier.

The inputs to a spectral clustering algorithm are a set of trajectories, say tr_1, $tr_2, \ldots, tr_n$, and a similarity matrix S, where any element s_{ij} of the matrix S denotes the similarity between trajectories tr_i and tr_j. It is to be noted that we consider $s_{ij} = s_{ji} \geq 0$, where $s_{ij} = 0$ if tr_i and tr_j are not similar. Given these two inputs, spectral clustering creates clusters of trajectories such that all trajectories in the same cluster are similar. In contrast, two trajectories belonging to different clusters are not so similar. For example, for the given set of trajectories, spectral clustering creates separate clusters for trajectories following different lanes or those that change lanes at the intersection. As a result, spectral clustering also helps identify the outliers and anomalous trajectories that are not similar to any other trajectories in the set. The two inputs to a spectral clustering algorithm may be represented as a graph $G = (V, E)$, where vertex v_i in this graph represents a trajectory tr_i. Two vertices are connected if the similarity s_{ij} between the corresponding trajectories tr_i and tr_j is greater than a certain threshold ϵ, and the edge is weighted by s_{ij}. Thus, the clustering problem may be recast as finding connected components in the graph such that the sum of the weights of edges between different components is negligible. Hence, by obtaining the

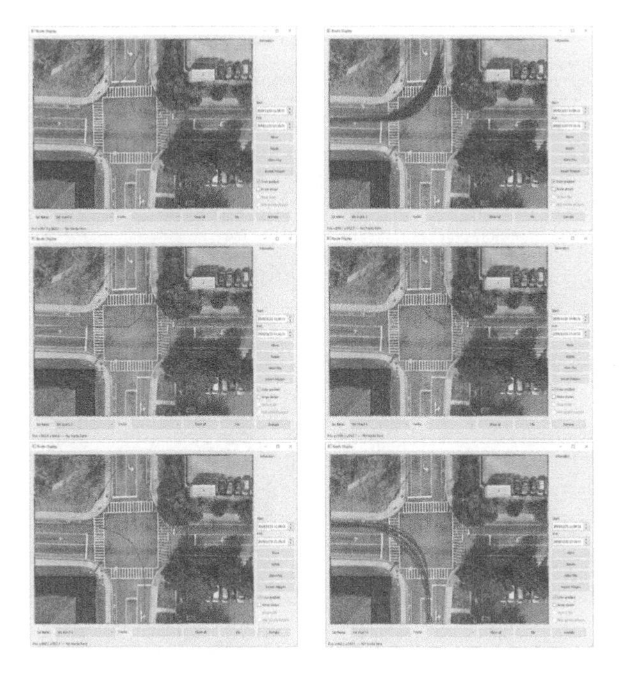

Figure 3.7 Second level of the two-level hierarchical clustering scheme where spectral clustering is applied to trajectories with the same direction of motion. The clusters of some left-turn trajectories are presented in this figure

number of connected components, we know the number of clusters and then get the clusters by applying a K-means algorithm.

We used the *linalg* library provided by NumPy to perform the linear algebra operations in spectral clustering. Once the number k of connected components is known, a K-means algorithm is run on the first k eigenvectors to generate the clustering results. The function $KMeans$ is a K-means clustering algorithm from the Python sklearn.cluster library.

The clusters generated by spectral clustering for all the left-turn trajectories are shown in Figure 3.7. The straight and right-turn trajectories clusters are omitted here due to space constraints.

3.5.2.3 Finding Representative Trajectories

After the trajectories are clustered, we identify a trajectory in each cluster that is representative of that cluster. The representative for each cluster is computed as the trajectory t belonging to the cluster with the least average distance from all the other trajectories.

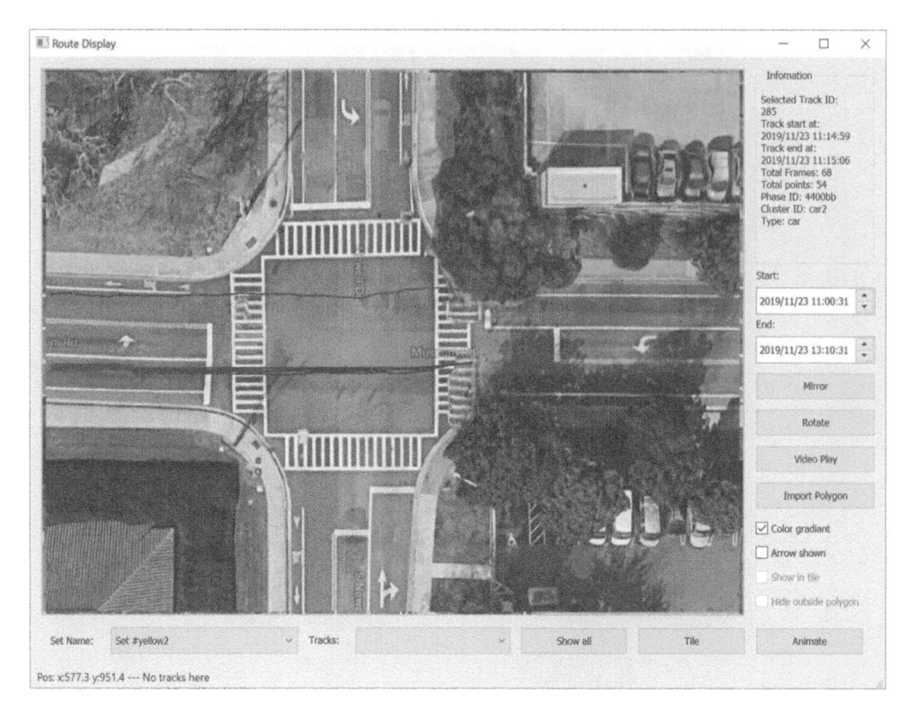

Figure 3.8 Collection of tracks representing vehicles that enter the intersection on a yellow light

3.6 ANOMALOUS BEHAVIOR

Detecting anomalous traffic behavior is one of the top goals for clustering trajectories. An anomalous trajectory may violate the spatial or temporal constraints at an intersection. The spatial constraints amount to the restrictions a vehicle must follow at an intersection, such as never going the wrong way. Temporal constraints, on the other hand, are the restrictions imposed by the signaling system at an intersection. We consider these two types of anomalous behavior in the rest of this section.

3.6.1 SIGNAL TIMING VIOLATIONS

The fusion of video and signal data allows us to detect the validity of the trajectories with reference to the current signaling phase of the intersection. The video and signal clocks are sometimes off by a few seconds. Adding an offset to the trajectories may treat the clocks as synchronous. This offset may be computed manually by comparing the time a signal in the video transitions to green and the time in the signaling when there is a "Phase Begin Green" event for the corresponding phase. It is also possible to compute the offset automatically in software by checking the timestamp of the first trajectory that crosses, say, the phase 2 stop bar (Figure 3.6) and the timestamp in signal data when the phase 2 signal becomes green and then adding 2.5 seconds of driver reaction time to the signal transition timestamp. Figure 3.8 shows the trajectories during a yellow light.

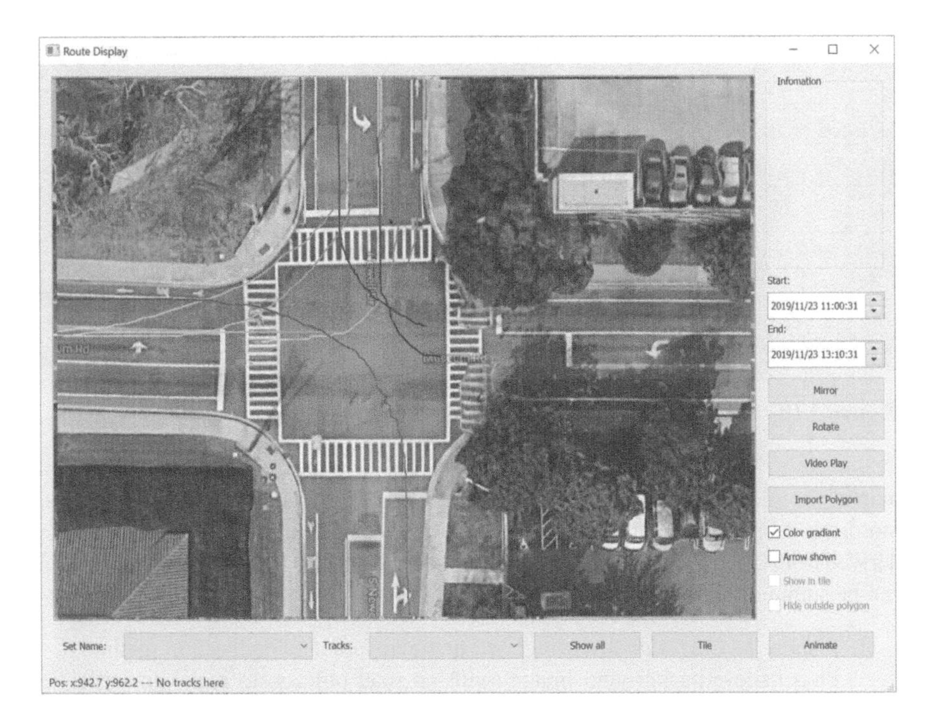

Figure 3.9 Collection of tracks representing vehicles with anomalous behavior because of their shape

3.6.2 TRAJECTORY SHAPE VIOLATION

Figure 3.9 shows the anomalous trajectories. In all the cases here, the trajectories are turn movements. Sometimes these trajectories take a very wide turn. At other times, the trajectories turn left from a through lane, and at still other times, the trajectories start taking a turn much before the actual stop bar, causing wrong-way access to the adjacent lane.

3.7 NEAR-MISS DETECTION FRAMEWORK

Our video processing software allows one to collect sufficient samples and visual cues corresponding to near-misses, intending to detect and even anticipate dangerous scenarios in real time so that appropriate preventive steps can be undertaken. In particular, we focus on near-miss problems from large-scale intersection videos collected from fisheye cameras (Figure 3.10). The goal is to temporally and spatially localize and recognize near-miss cases from the fisheye video. The primary motivation for resolving distortion instead of using original fisheye videos is to compute accurate distance among objects and their proper speeds using rectangular coordinates that better represent the real world. The projections are made on an overhead satellite

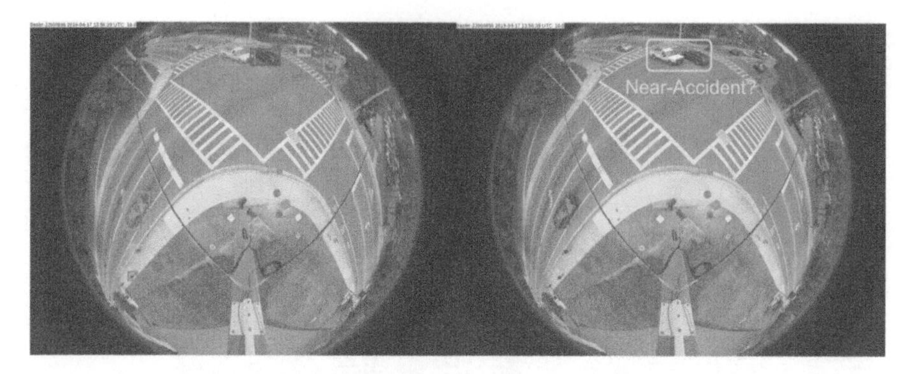

Figure 3.10 An illustration of the near-miss detection problem

map of the intersection. We specify five categories of objects of interest–pedestrians, motorbikes, cars, buses, and trucks. The overhead satellite maps of intersections are derived from Google Earth®. The main steps of our detection framework (Figure 3.11) can be summarized as follows:

1. *Fisheye to Cartesian Mapping:* We first apply camera calibration methods on a fisheye background image (with no road objects) to make an initial correction. We take the calibrated image as the target image and an overhead satellite map as the reference image and select corresponding landmark points in both images for mapping. Given these landmark points, we adopt the thin-plate spline (TPS) [71, 72] as the basis function for coordinate mappings from the reference to the target and store the point-to-point outputs.
2. *Object Detection and Multiple Object Tracking:* We train an object detector using deep learning techniques and design a vehicle reidentification model with deep cosine metric learning to handle occlusion problems. We integrate these two models into our multiple-object tracking pipeline. The framework

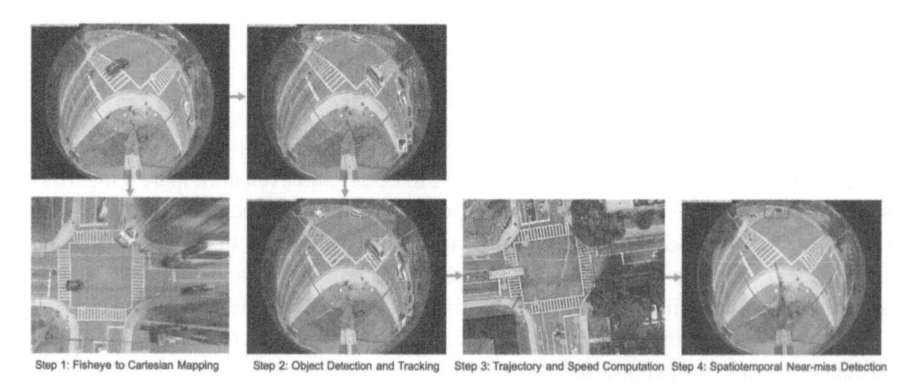

Figure 3.11 The pipeline overview of the proposed framework

supports real-time object detection and multiple object tracking. For more details on object reidentification using cosine metric learning, refer to Sections 2.3.3 and 2.4.5.2.

3. *Trajectory and Speed Computation:* Using the point-to-point TPS mappings, we correct and scale road object trajectories and speed information from the perspective of the overhead satellite map with learned deep features. The complexity of coordinates transfer is $O(1)$, allowing us to process online and offline data.
4. *Spatial and Temporal Near-miss Detection* We define two scenarios for near-misses in videos: (1) spatial scenario–proximity of road objects in image space; (2) temporal scenario – a dramatic speed decrease to avoid near-misses (a sudden brake). We use distance-based and speed-based measures to compute the near-miss probabilities of road objects and aggregate scores via averaging as the final output.

Figure 3.12 demonstrates the pipeline and the overall architecture of the proposed method.

3.7.1 FISHEYE TO CARTESIAN MAPPING

3.7.1.1 Calibration and Perspective Correction

Due to fisheye lens distortion and perspective distortion, we found that directly applying mapping methods between fisheye images and satellite maps does not result in good-quality mappings. Therefore, we wish to utilize fisheye camera parameters to make an initial calibration. For our fisheye camera model, points in a real 3D world are first transformed to fisheye coordinates via extrinsic parameters (rotation and translation). These fisheye coordinates are mapped into the 2D image plane via the

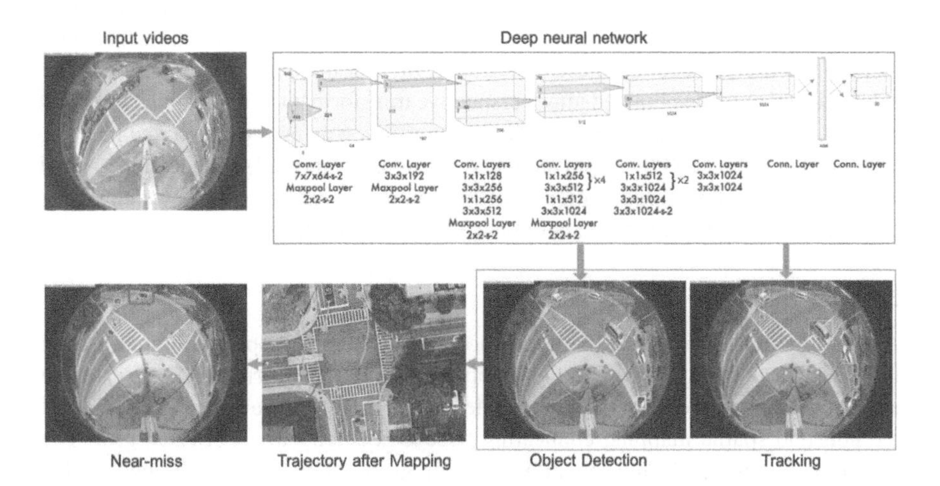

Figure 3.12 The pipeline and the deep model architecture of the proposed method

intrinsic parameters (including the polynomial mapping coefficients of the projection function). For a point, P, in the 3D world, the transformation from world coordinates to points in the camera reference image is as follows:

$$\begin{pmatrix} x \\ y \\ z \end{pmatrix} = \begin{pmatrix} Xc \\ Yc \\ Zc \end{pmatrix} = R \begin{pmatrix} Xw \\ Yw \\ Zw \end{pmatrix} + T, \tag{3.4}$$

where R is the rotation matrix and T stands for translation. The pinhole projection coordinates of P are (a, b) where $a = x/z$, $b = y/z$, $r^2 = a^2 + b^2$, $\theta = atan(r)$. The fisheye distortion is defined as

$$\theta_{distortion} = \theta(1 + k_1\theta^2 + k_2\theta^4 + k_3\theta^6 + k_4\theta^8), \tag{3.5}$$

where the vector of distortion coefficients is (k_1, k_2, k_3, k_4) and camera matrix is

$$A = \begin{bmatrix} f_x & 0 & c_x \\ 0 & f_y & c_y \\ 0 & 0 & 1 \end{bmatrix} \tag{3.6}$$

The distorted point coordinates are $(x' = (\theta_d/r)a, y' = (\theta_d/r)b)$. The final pixel coordinates vector is (u, v), where $u = f_x(x' + \alpha y') + c_x$ and $v = f_y y' + c_y$, and skew coefficient α is set to zero and remains zero.

The distortion correction procedure involves three major stages: calibration correction, perspective correction, and TPS mapping. The calibration process consists of getting parameters using a checkerboard reference. The image obtained after the calibration has a noticeable perspective distortion, which is adjusted by selecting four points in the output image of the first stage and then mapping them to a reference satellite image. There are small but noticeable distortions in the image after perspective correction caused by the structural elements of the road, such as ridges and grooves, or they may be due to minor errors caused in the calibration. The TPS mapping is used to address these distortions. In TPS mapping, multiple points are selected on the image obtained after perspective transformation and mapped to points on the satellite map. It approximates the transformation using a spline-based method. Thus, by performing TPS, we get an image whose ground (road) and the map ground almost overlap. As our application goes beyond distortion correction, we can track the vehicles and get the exact location in cartesian coordinates.

3.7.1.2 Thin-plate Spline Mapping

After calibration and perspective correction steps, we can compute an initial fisheye to cartesian mapping. To refine the mapping between the corrected fisheye image and satellite map, we adopt the thin-plate spline (TPS) as the parameterization of the nonrigid spatial mapping connecting fisheye geometry to a Cartesian grid. The choice of TPS to handle the spatial warping in our problem is driven by the fact that it is a natural nonrigid extension of the affine map. Furthermore, we do not have any information regarding physics-based mappings that can augment fisheye calibration.

Figure 3.13 An illustration of omnidirectional fisheye camera used for data collection and examples of fisheye video

Therefore, we adopt the TPS to generate mappings. Given the point sets V and Y in 2D ($D = 2$) consisting of points $v_a, a = 1, 2, ..., K$, and $y_a, a = 1, 2, ..., N$, respectively, the TPS fits a mapping function $f(x, y)$ using corresponding landmark sets y_a and v_a by minimizing the following energy function [72]:

$$\begin{aligned} E_{TPS}(f) &= \sum_{a=1}^{K} \|y_a - f(v_a)\|^2 \\ &+ \lambda \int\int \left[(\frac{\partial^2 f}{\partial x^2})^2 + 2(\frac{\partial^2 f}{\partial x \partial y})^2 + (\frac{\partial^2 f}{\partial y^2})^2 \right] dxdy. \end{aligned} \tag{3.7}$$

Homogeneous coordinates are used for the landmarks with each point y_a represented as a vector $(1, y_{ax}, y_{ay})$. With a fixed regularization parameter λ, a unique minimizer, f, can be obtained as follows [72]:

$$f(v_a, d, w) = v_a \cdot d + \phi(v_a) \cdot w, \tag{3.8}$$

where d is a $(D + 1) \times (D + 1)$ matrix representing the affine transformation and w is a $K \times (D + 1)$ warping coefficient matrix representing the nonaffine deformation. The vector $\phi(v_a)$ is a $1 \times K$ vector related to the TPS kernel. When combined with the warping coefficients w, the TPS generates a nonrigid warping. Figure 3.13 illustrates the omnidirectional fisheye camera model, camera placement, and examples of collected video data. We present the mapping results of two intersections in Figure 3.14.

3.7.2 TRAJECTORY AND SPEED COMPUTATION

We leverage tracking results to generate a trajectory for each object in terms of frame, track ID, class, and $x - y$ coordinates. We transform the $x - y$ coordinates from fisheye

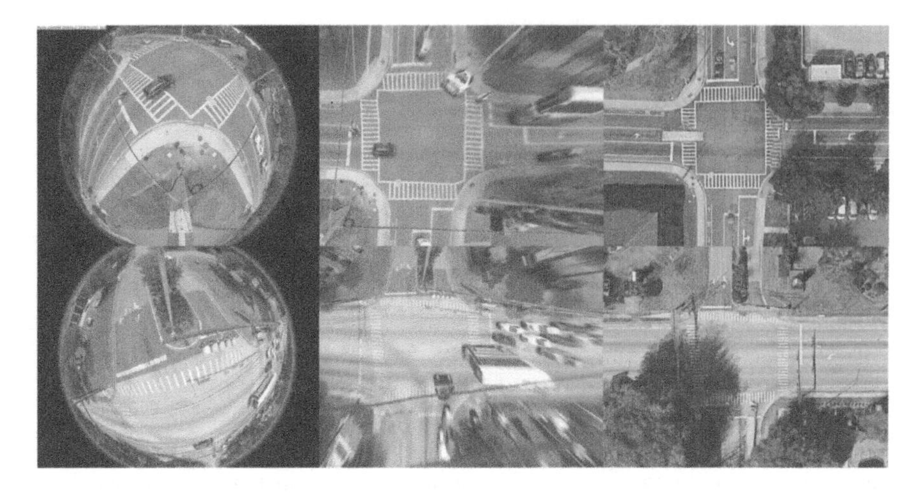

Figure 3.14 Calibration and TPS mapping are used for fisheye to cartesian mapping. *Left:* Original fisheye image. *Middle:* Mapping result. *Right:* Reference satellite map

image space to overhead satellite map space using the point-to-point mapping matrix obtained in the mapping pipeline. We estimate the speed of objects using distance after mapping. To leverage more accurate and compact object masks than rectilinear bounding boxes, we also investigate Huang et al.'s [73] use of gSLICr [22], a GPU-based implementation of SLIC [57] – a superpixel segmentation method – instead of standard rectangular bounding boxes. Figure 3.15 demonstrates the use of superpixels for generating object masks. This integration performed in real time results in better distance measures that can be utilized for detecting near-misses.

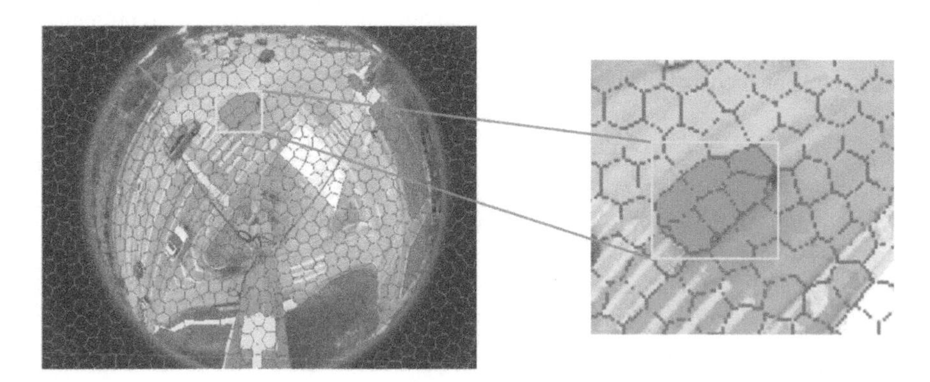

Figure 3.15 An example of superpixel segmentation on fisheye video. It is used for extracting more detailed object features (with object boundaries and shapes) than detection

3.7.3 NEAR-MISS DETECTION

Our method performs object detection and multiple object tracking in real time and can handle large-scale and city-scale intersection videos for traffic understanding and analysis. Using our TPS-based nonrigid mapping tool, we can correct online and offline coordinates to project road object locations to satellite maps and then form refined trajectories for near-miss detection.

Two near-miss scenarios are defined for videos: (1) spatial scenario – road objects collide or are very close in image space, and (2) temporal scenario – a dramatic speed decrease to avoid near-misses (a sudden brake). We use distance-based and speed measures to compute the near-miss probability of road objects with an average of two scores as the final output described below.

Spatial Distance Measure. We use track data to form trajectories of road objects and compute distances between two road objects using center coordinates of detected bounding boxes in image space at the frame level. The probability of a spatial near-miss is calculated using the Euclidean distance of road objects with a ratio to the size of the object according to object class (vehicle size of each class does not vary much) and is computed as follows:

$$P_{spatial}(\mathbf{b}_t^p, \mathbf{b}_t^q) = \frac{1}{2} \cdot \frac{(\mathbf{w}_t^p + \mathbf{w}_t^q + \mathbf{h}_t^p + \mathbf{h}_t^q)}{\sqrt{(\mathbf{x}_t^p - \mathbf{x}_t^q)^2 + (\mathbf{y}_t^p - \mathbf{y}_t^q)^2}} \tag{3.9}$$

where $\mathbf{b}_t^p$ and $\mathbf{b}_t^q$ denote the detected bounding boxes for the pth and qth objects in the tth frame. $\mathbf{w}_t^p$, $\mathbf{h}_t^p$, $\mathbf{x}_t^p$, $\mathbf{y}_t^p$ denote the object width, object height, x coordinate, and y coordinate for the pth object in the t-th frame, respectively.

Temporal Motion Measure. The speed of the road object is computed by adjacent displacement over multiple time frames. The probability of motion-based near-miss is computed by the fractional decrease in speed and is computed as follows:

$$P_{temporal}(\mathbf{b}_{k:t}^p) = \frac{\max \sum_{i=k}^{t} (\mathbf{s}_{i+1}^p - \mathbf{s}_i^p)}{average(\mathbf{s}_{k:t}^p)} \tag{3.10}$$

where $\mathbf{b}_{1:t}^p$ denotes the detected bounding boxes for the pth from its first frame (kth frame) to its last frame (tth frame). $\mathbf{s}_i^p$ denotes speed for the pth object in the ith frame. We use a weighted average of the above two probabilities of near-miss to compute the overall score.

3.8 EXPERIMENTS

We first describe the dataset used for our experimental evaluation. We then present qualitative performance and quantitative assessment of our methods for object detection, multiple object tracking, superpixel segmentation, thin-plate spline, and near-miss detection. We present a performance comparison between a nonmapping-based method and our proposed calibration+TPS-based method for near-miss detection.

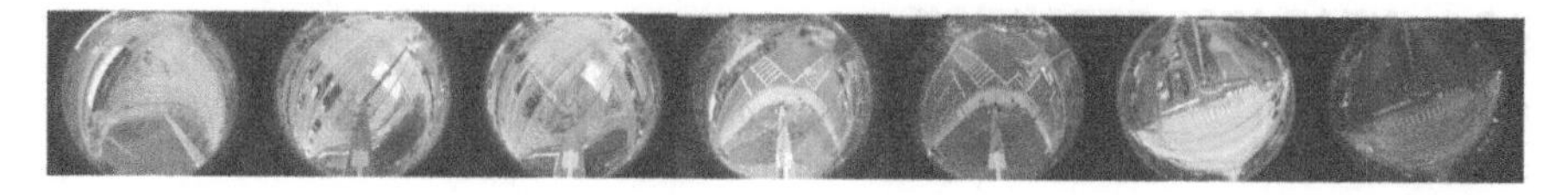

Figure 3.16 Gallery images of fisheye video with various locations (four cameras) and lighting conditions

3.8.1 FISHEYE VIDEO DATA

We have collected large-scale fisheye traffic video from omnidirectional cameras at several intersections. Figure 3.16 shows gallery images of the dataset with several collected fisheye video samples at multiple intersections under different lighting conditions. We collected 8 hours of videos daily for each intersection: 2 hours each for the morning, noon, afternoon, and evening. The total video datasets used in the experiments have a duration of more than 100 hours. As discussed earlier, fisheye intersection videos are more challenging than videos in other datasets collected by surveillance cameras for reasons including fisheye distortion, multiple object types (pedestrians and vehicles), and diverse lighting conditions. We manually annotated the spatial location (bounding boxes) and temporal location (frames) for each object and near-miss to generate ground truth for object detection, tracking, and near-miss detection.

3.8.2 QUALITATIVE PERFORMANCE

3.8.2.1 Fisheye to Cartesian Mapping

The results for the calibration+TPS pipeline (Figure 3.14) show that fisheye distortion and perspective distortion are effectively addressed by our method. The qualitative results in terms of performance for object detection, multiple object tracking, and superpixel segmentation (Figure 3.17) show that the deep learning–based detector is effective in classifying objects even when the image footprint is small (e.g., pedestrians and motorbikes). The use of deep cosine metric learning allows the tracker to generate more consistent and stable tracks. The superpixel segmentation assists in outputting compact contours of objects. The latter can then be used for an effective signature for tracking.

3.8.2.2 Trajectory and Near-miss Detection

The trajectories of road objects projected on the satellite map, along with referenced tracking frames, are shown in Figure 3.18. These trajectory maps give an easier-to-understand traffic pattern for the intersection than that from the perspective of the original fisheye camera. Samples of near-misses that we detected at different intersections are presented in Figure 3.19. The first example shows a spatial near-miss between two road objects. The second example shows a temporal near-miss as the front white car suddenly stopped in the middle of the intersection, forcing a sudden brake for the following vehicle.

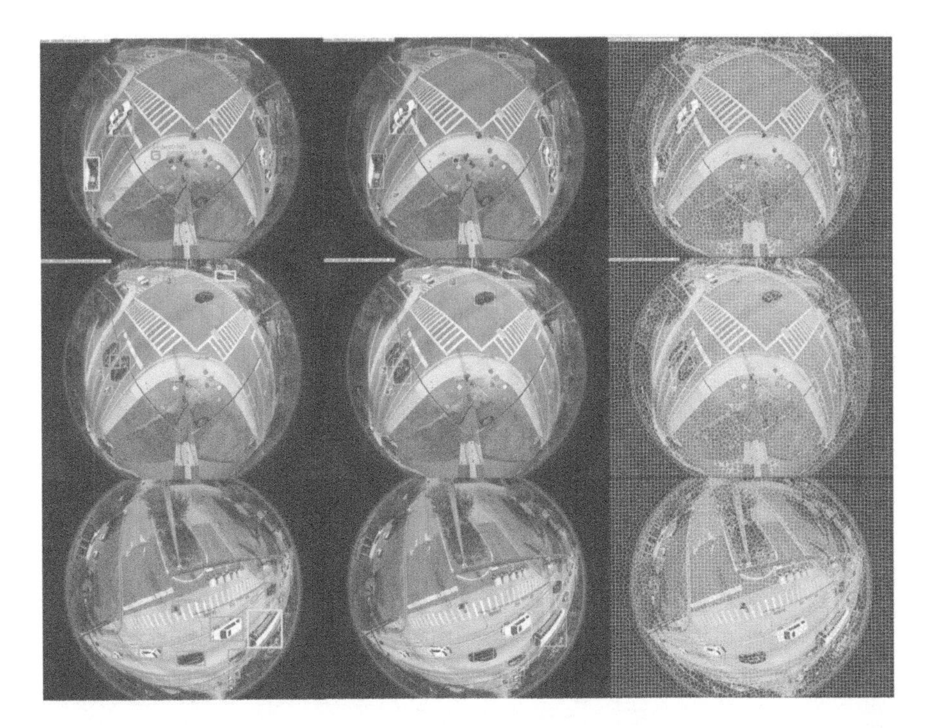

Figure 3.17 Qualitative results of detection, tracking, and segmentation tasks. *Left:* Object detection outputs the object class (car, pedestrian, bus, motorbike, etc.) and localization (bounding box). *Middle:* Multiple object tracking associates the object in consecutive video frames (track ID). *Right :* Superpixel segmentation aids in computing object boundaries and shapes

3.8.3 QUANTITATIVE EVALUATION

We present a quantitative evaluation of the overall performance of our proposed method in terms of speed performance, improvement of object speed measures based on mapping, and the precision and recall for each subtask of the pipeline.

3.8.3.1 Computational Requirements

We present the speed performance for the tested methods in Table 3.1. The fisheye video resolution is 1,280 $\times$ 960, and our implementation for thin-plate spline takes 10 seconds for one-to-one corresponding mapping for 1,228,800 points. This is a one-time setup cost.

After getting mapping point sets, all video processing experiments were performed on a single GPU (NVIDIA Titan V). The GPU-based SLIC segmentation [57] has excellent speed and can process 400 fps on fisheye videos. The overall pipeline of our methods (object detection, multiple object tracking, and near-miss detection) achieves about 40 fps. This rate is sufficient to address a variety of real traffic surveillance and near-miss detection for large-scale daily video data.

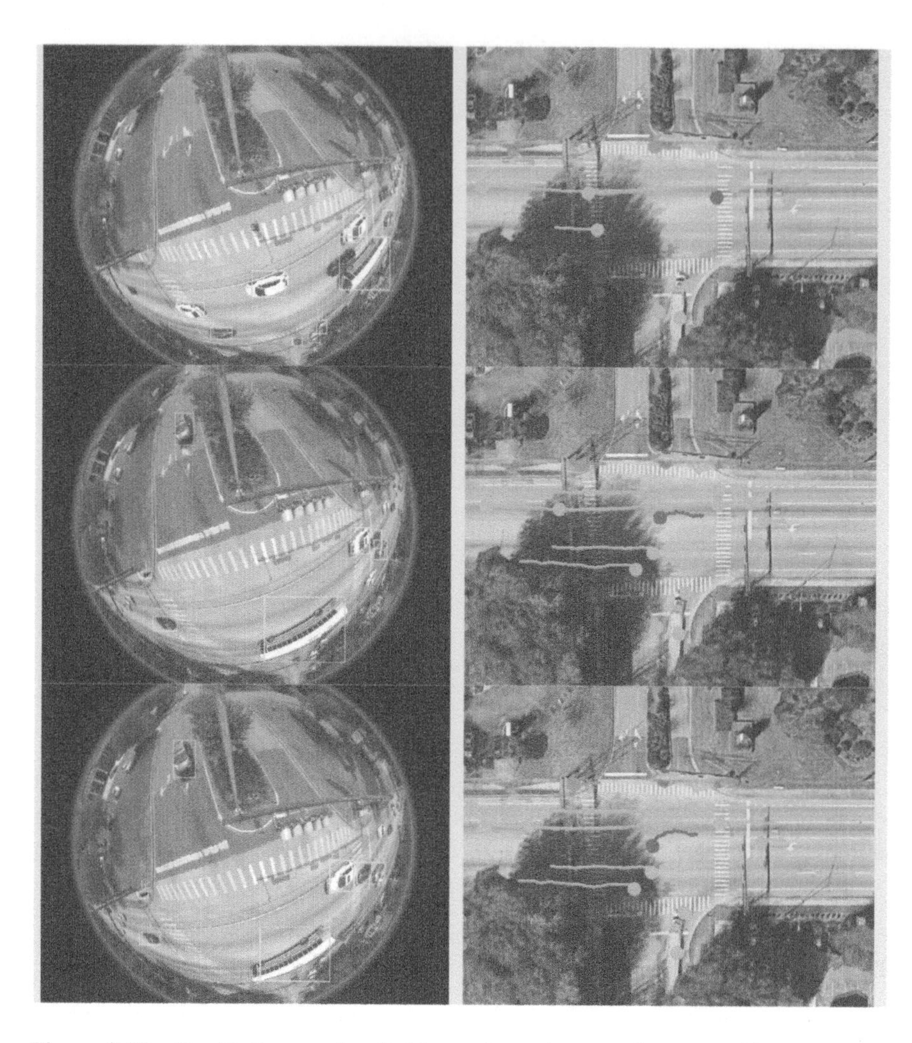

Figure 3.18 Qualitative results of object trajectories mapping to satellite map. *Left:* Tracking in fisheye video. *Right:* Trajectories after mapping. Different color represents different object class (red for bus, green for pedestrian, and blue for car)

3.8.3.2 Trajectory and Near-miss Detection

A quantitative prediction of object representation and near-miss is achieved by comparing predicted detection with the ground truth at the frame level. A true positive corresponds to a high level of overlap between the prediction and ground truth detection pair. It is computed using an Intersection over Union (IoU) score. The track is correctly associated if this overlap exceeds a predefined threshold (e.g., 0.7). A true negative means no prediction and no associated ground truth. A false positive is that a prediction has no associated ground truth. A false negative is that a ground truth

Figure 3.19 Qualitative results of two types of near-miss detected. *Top 3 images:* A spatial near-miss case – a motorbike and a car collide. *Bottom 3 images:* A temporal near-miss case caused a sudden brake

has no associated prediction. The true negative rate (TNR) also refers to specificity, and the false positive rate (FPR) refers to fallout. The specificity, fallout, precision, recall, and F1 score are defined as follows:

$$TNR = \frac{TN}{TN + FP} = 1 - FPR. \tag{3.11}$$

$$Precision = \frac{TP}{TP + FP}; \quad Recall = \frac{TP}{TP + FN}. \tag{3.12}$$

$$F1 = 2 \times \frac{Precision * Recall}{Precision + Recall}. \tag{3.13}$$

We compute object speed information based on trajectories by converting pixels to actual meters and frame intervals to seconds. Figure 3.20 shows an example of the comparison of computed object speed information where a car approaches the intersection with speed decreasing from 60 km/h to 20 km/h and then back to 30 km/h. With nonmapping methods, object speed computing suffers from fisheye and perspective distortion and yields inaccurate results. We also present accuracy evaluation for object and multiple object detection (cosine metric learning) in Table 3.2.

Table 3.1
Quantitative Evaluation of Speed Performance

Methods	CPU/GPU	Speed
TPS mapping	CPU	10 s
SLIC segmentation	Nvidia Titan V	400 fps
Overall pipeline	Nvidia Titan V	40 fps

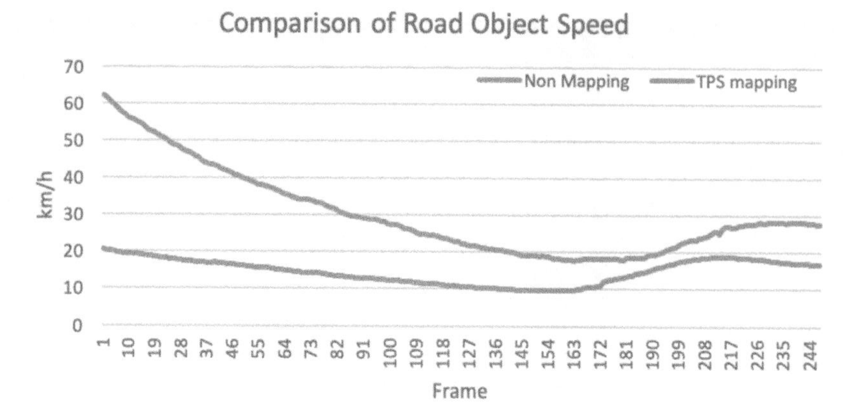

Figure 3.20 Quantitative comparison of computed object speed between non-mapping and proposed methods

Table 3.2

Quantitative Performance of Object Detection and Multiple Object Tracking

Methods	TP	FN	FP	Precision	Recall	F1 Score
Object detection	7,649	102	82	0.98940	0.98684	0.98812
Multiple object tracking	7,540	483	314	0.96002	0.93980	0.94980

Table 3.3

Quantitative Comparison of Near-miss Detection Between Nonmapping and Calibration+TPS-based Method

Methods	Intersection	TN	FP	Specificity	Fallout
Nonmapping based (baseline)	Intersection 01	2,869	32	0.98897	0.01103
	Intersection 02	2,659	162	0.94257	0.05743
Calibration + TPS mapping	Intersection 01	2,895	6	0.99793	0.00207
	Intersection 02	2,818	3	0.99894	0.00106

As real near-miss is rare in terms of two-camera video data in a week, it is more reasonable to examine specificity (selectivity or TNR) and fallout (FPR) for near-miss detection. In Table 3.3, we present the comparison of non-mapping-based detection and calibration+TPS mapping-based detection in terms of TNR and FPR. The quantitative evaluation demonstrates the overall effectiveness of our proposed method for near-miss detection in large-scale fisheye traffic videos.

3.9 DISCUSSION

Using a deep learning model integrated with camera calibration and spline-based mapping methods, we presented a novel unsupervised approach to detect near-misses in fisheye intersection video. It maps road object coordinates in fisheye images to a satellite-based overhead map to correct fisheye lens and camera perspective distortion. This allows for computing distance and speed more accurately. This unified approach performs real-time object recognition, multiple object tracking, and near-miss detection in fisheye video. Handling geometry on object-level analysis in the fisheye video is efficient and robust, resulting in more accurate near-miss detection. The experimental results demonstrate the effectiveness of our approach, and we show a promising pipeline broadly applicable to fisheye–video-understanding applications such as accident anticipation, anomaly detection, and trajectory prediction.

As we get rich spatial and temporal features of road objects in both fisheye image space and overhead satellite image space, these track data can be easily applied to several other traffic and computer vision tasks such as anomaly detection, trajectory clustering, trajectory prediction, turn movement counts, etc. Intersection signal data can be integrated with video data to develop interesting traffic analyses, e.g., cars crossing the intersection during a red light. The generated tracks can be plotted over extended periods to visualize macro-trends.

4 Severe Events

4.1 INTRODUCTION

Intersection safety is an active area of research because traffic intersections are prone to crashes. USDOT estimates more than 50% of road crashes leading to fatality or injury happen at or near traffic intersections. Road crashes have been one of the leading causes of death worldwide. With the rapid advancement in technology, many intersections now have video cameras deployed as sensors to monitor these intersections. The videos are streamed over the web, stored, and processed for safety assessment. Existing intersection safety assessment methodologies often require the analysis of historical data to infer current and future intersection user behavior. Although helpful, these data are often biased to what has been reported, incomplete, and retrospective. This chapter will present our end-to-end intersection safety methodology, starting with processing intersection videos and computing existing and new "surrogate safety measures."

Based on decades of safety research using crash data, it is generally acknowledged that using surrogate safety measures could provide further insights into enhancing the safety of roadways. These surrogate measures rely on maneuvers (trajectories) of vehicles and pedestrians. Countermeasures can be developed to reduce or eliminate unsafe maneuvers by understanding trajectories that could have led to a crash. The most common surrogate safety measure is the "near-miss" or the "traffic conflict." Near misses involve a vehicle's trajectory coming very close to that of another vehicle or a pedestrian without an actual collision. The proximity of the trajectories is measured on a temporal scale using metrics such as time-to-collision (TTC) [74], and post-encroachment time (PET) [75]. The severity of the near-miss event can be determined based on the temporal proximity measures (TTC and PET) and the velocities of the vehicles or pedestrians involved. Typically, shorter TTC and PET, combined with higher relative speeds, imply a greater severity of the near-miss (i.e., the lesser likelihood of not avoiding the potential crash and higher injury severity had the crash not been avoided). Unlike the case of crashes, there are currently no clear thresholds or categories for classifying near-misses by severity. In this chapter, we introduce a new surrogate safety measure, "severe event," that includes all near-miss events and unsafe behavior exhibited by road users.

While surrogate safety measures offer an excellent opportunity to understand site-specific and time-specific safety issues and develop countermeasures, a tremendous practical impediment in using surrogate measures for safety analysis is the need to process large volumes of video data to determine trajectories, identify the conflicts in these trajectories, and filter these down to a subset of critical unsafe maneuvers for further analysis. In signalized intersections, it is also necessary to analyze the unsafe maneuvers for the ongoing signal phasing data to identify the appropriate countermeasures. For example, unsafe maneuvers during a permitted left-turn phase

DOI: 10.1201/9781003431176-4

may suggest the need for a protected left-turn phase. Similarly, conflicts between right-turning vehicles and crossing pedestrians may suggest separating the signal phases for these two movements (no right turn on red or leading pedestrian phase). This chapter makes several contributions to the field of intersection safety analysis, described as follows:

1. The algorithms to decompose pedestrian and vehicle trajectories while fusing signal timing data to derive features useful for safety analysis.
2. Introduction of a new surrogate safety measure, severe event, quantified by multiple existing metrics such as TTC or PET as recorded in the event, deceleration, and speed.
3. Categorization of severe events based on the directional movement of vehicles, pedestrian movements, and phase information. This is then used for the relative weighting of severe events (based on the potential for damage) and deriving a weighted average to reflect a comparative safety rating for each intersection. This comparison can also consider the exposure rate (determined by the number of vehicles and pedestrians crossing the intersection).
4. An efficient multistage event filtering approach followed by a multi-attribute decision tree approach that subsets the extensive set of conflicting interactions to a robust set of severe events.

We have applied our algorithms to multiple intersections in two different cities. Our extensive results demonstrate the usefulness of the software by providing key insights into severe conflicts by the day-of-the-week and hour-of-the-day analysis.

The rest of the chapter is organized as follows. Section 4.2 presents the related work in traffic safety analysis using surrogate safety measures. Section 4.3 presents the overall methodology we developed, starting with trajectory generation and computation of features from the trajectories, our strategy of classification of serious conflicts, and the introduction of the new surrogate safety measure, severe event that we developed as a part of this work. We present a set of event filters to filter out events that are not potentially dangerous automatically. Section 4.4 presents our experimental results, and we conclude in Section 4.5.

4.2 RELATED WORK

Several surrogate safety measures have been developed, relying on the physical properties (time, distance, and speed) of vehicle trajectories. Measures such as TTC and PET, which are based on the temporal proximity of the road users, are perhaps the most widely used indicators, especially in the context of intersections, which are the focus of this chapter. TTC is the time remaining to avoid a collision, from when the road user takes action to where the collision can occur [74]. PET is the time difference between when the first road user leaves one point and when the second road user arrives at that same point [75]. Lower values of TTC and PET indicate higher risks of collision. Several other measures have been proposed based on either spatial proximity or acceleration–deceleration patterns of vehicles [76, 77]. A very comprehensive synthesis of the literature on surrogate safety measures was recently provided by Arun et al. [78].

Just as all traffic crashes are not equally severe (some could lead to fatalities while a vast majority are minor crashes with only property damage but no injuries), all temporally proximal interactions among road users need not be "safety-critical" events. Even though Hydén [79] proposed over three decades ago that there is a hierarchy of traffic events varying in severity, Arun et al. [78] note that there is still no consensus on what constitutes a safety-critical event or a near-miss.

One approach to identifying critical events is applying thresholds on surrogate safety measures. For example, thresholds on TTC range from 1.5 seconds to 3.0 seconds [80, 81], while those for PET range from 1.0 second to 1.5 seconds sec [82]. Broadly, the perception-reaction time of road users (the time taken by a road user to understand a situation and react to it) is considered a benchmark in determining these thresholds.

Surrogate safety measures primarily reflect the possible interactions (or "events") between road users; not all are critical from a safety standpoint. Therefore, it is essential to distinguish between safer and critical interactions. Safety-critical events are also known as near-misses or sometimes as traffic conflicts, but the definition of a traffic conflict has remained contentious over years [78]. The early definitions of traffic conflicts indicate that only the most extreme traffic interactions have been considered safety-critical. Hydén [79] proposed a hierarchy of traffic events varying in severity. Arun et al. [78] note that there is still no consensus on what constitutes a safety-critical event or a near-miss.

A second approach to identifying critical conflicts is the Swedish Traffic Conflict Technique [83]. This approach considers both the surrogate measure and the speed of the conflicting road users. In general, events representing a combination of lower times to crash and a higher conflict speed are considered more serious events. Figure 4.1 presents an example of conflict curves. Vehicle–vehicle conflicts above curve 26 are considered serious, while vehicle–pedestrian conflicts above curve 24 are considered serious [84]. Determining surrogate safety measures and critical safety events requires trajectories of road users (vehicles and pedestrians) as inputs. These trajectories may be obtained from traffic simulators or the processing of real-world video data. The Surrogate Safety Assessment Model (SSAM) has been developed as a post-processor to estimate the number and severity of conflicts based on vehicle trajectory data [84]. The outputs of SSAM include the number, the type, the severity, and the locations of three types of simulated conflicts (crossing, lane changing, and rear-end). The conflict type is identified according to the lane and link information or the angle between the two converging vehicles.

Commercial applications for processing trajectory data from videos to determine surrogate measures include those developed by DERQ, AMAG, Currux, and Transoft Solutions. Our work in this chapter is closely related to McLauchlan et al. [85], where the authors use video analysis for intersection traffic analysis. The main difference is that Ref. [85] has not treated pedestrian–vehicle conflicts. Further, only PET was used as a safety indicator. In contrast, we use a holistic approach to compute and use

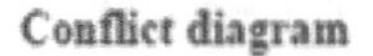

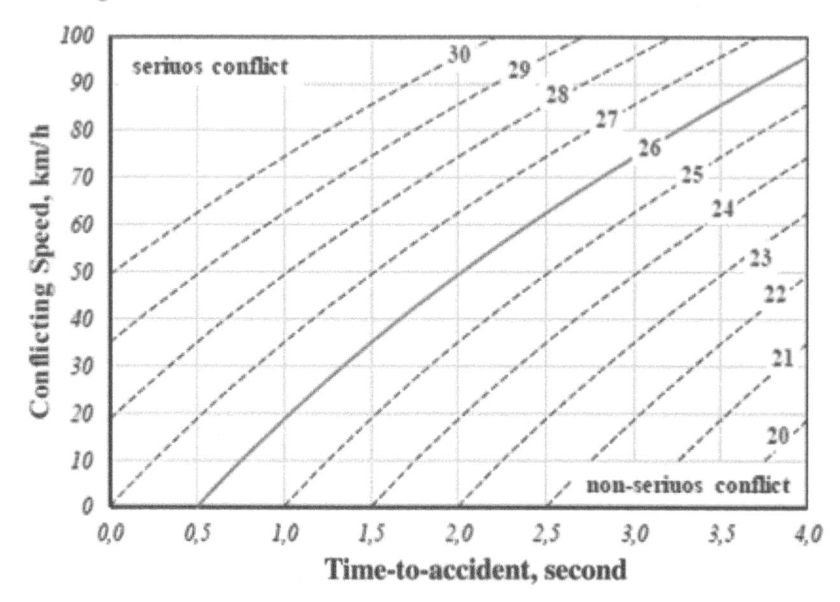

Figure 4.1 Swedish Technique Conflict diagram (reproduced from Laureshyn and Várhelyi [83])

PET and TTC (as appropriate), speed, deceleration, and distance between the road users as active features for severe event detection. These features give a deeper insight into the event's nature and get us nearer to the events that are indeed close calls.

4.3 METHODOLOGY

The methodology used for this work is described as follows. Section 4.3.1 presents the salient features generated by our software to qualify the events and their use for safety analysis. Section 4.3.2 introduces a new surrogate safety measure, severe event, and describes a categorization scheme for vehicle–vehicle and pedestrian–vehicle events. Section 4.3.3 describes a filtering process, an event sieve that helps narrow down the events to keep only the most critical events. Finally, Section 4.3.4 describes a multi-attribute decision tree approach to isolate high-intensity regions in feature space containing the most severe events.

4.3.1 GENERATING FEATURES FROM TRAJECTORIES

Trajectories are generated by processing the intersection videos. The safety analysis system takes fisheye video footage as input, then annotates objects with bounding boxes, maps those coordinates from the fisheye image to rectilinear space, and then stores the results in the trajectory table. The object detection and tracking module utilizes YOLOv4 [86] to detect different kinds of road participants, including vehicles, pedestrians, cyclists, and motorcyclists. A modified DeepSORT algorithm associates detections across frames and assigns a unique ID for each object. As the trajectories from fisheye videos are usually of unnatural shapes caused by lens distortion, we perform rectification and alignment to Google Maps images before feeding the trajectories to downstream modules. Our solution for rectification includes two steps: fisheye-to-perspective transformation followed by thin-plate spline (TPS) warping.

Separately, the city's Automated Traffic Signal Performance Measures (ATSPM) data are collected, which provides the signal information for each traffic light per intersection over time. This signal information is merged with object trajectories, enabling analyses concerned with object location and signal states over time, such as signal violations, lingering mid-trajectory, etc. However, one issue that arises is the synchronization of city-provided and video-recorded signal changes, which usually vary by a few seconds. A purpose-built computer vision model is trained to output signal states, which are compared to ATSPM signal changes to yield the time delay between video and city data. If a signal is not visible, then the start-up time of a vehicle is used under the assumption that driver reaction time is roughly 1.5 seconds, but this time is configurable in the software.

We have computed a comprehensive set of features for every conflict event. A feature in this context is an individual measurable property or characteristic of a conflicting event. The following is a list of the key features that we compute:

1. *Standard Near-miss Attributes:* We compute the common risk assessment metrics, such as TTC and PET, for every event.
2. *Signal-phase Information:* The fused video-ATSPM dataset is used to determine features such as the ongoing vehicle signal, ongoing pedestrian signal, and if the event occurs during the beginning, middle, or end of the current signaling phase.
3. *Trajectory Features:* The trajectory-related features are the trajectory's movement, phases, and lanes.
4. *Speed Features:* These include the current speeds and accelerations for vehicle-vehicle interactions.
5. *Distance:* Spatial distance between two users at the time of the conflict.

Figure 4.2 shows the intersection labels for the crosswalks that will be used throughout the book.

4.3.2 CATEGORIZATION OF SEVERE EVENTS

The categorization of the vehicle–vehicle and pedestrian–vehicle conflicts is described in this section.

1. *Pedestrian–Vehicle (P2V) Events:* The following are the main conflict events between vehicles and pedestrians at signalized intersections:

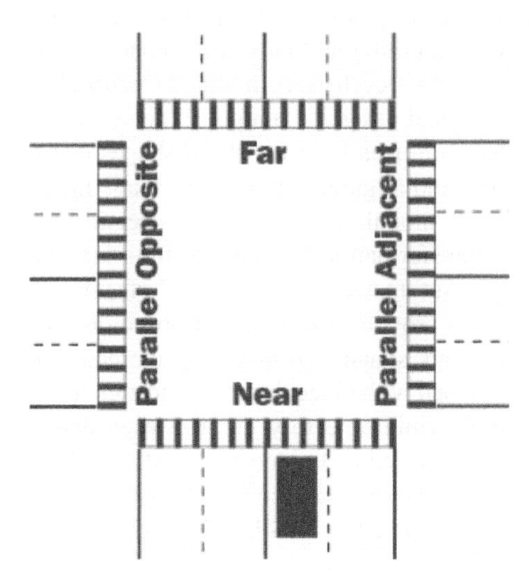

Figure 4.2 Crosswalk naming convention with respect to the vehicle

- *Conflict Types 1 and 2:* Right-turning vehicle with the pedestrian in an adjacent parallel crosswalk (Figure 4.3a) and near crosswalk (Figure 4.3d), respectively.
- *Conflict Types 3 and 4:* Left-turning vehicle with the pedestrian in the parallel opposite crosswalk (Figure 4.3b) and adjacent parallel crosswalk (Figure 4.3d), respectively.
- *Conflict Types 5 and 6:* Through vehicle with the pedestrian in the far crosswalk (Figure 4.3c) and the near crosswalk (Figure 4.3d), respectively.

Among these conflicts, Conflict Types 1 and 3 are possible conflicts between pedestrians and vehicles at signalized intersections if all vehicles and pedestrians strictly follow the traffic rules (assuming a three- or four-leg intersection, which are the most common).

2. *Vehicle–vehicle (V2V) Events:* As the primary purpose of signalization is to reduce or eliminate conflicting movements on the intersection, the following are the possible conflicts between vehicles at signalized intersections if all vehicles strictly follow the traffic rules (assuming a three- or four-leg intersection, which are the most common):
   - *Left turn and Opposing Through (LOT):* A left-turning vehicle in a permitted phase conflicts with an opposing through movement (Figure 4.4a).
   - *U-turn and Opposing Through:* A U-turning vehicle in a permitted phase conflicts with an opposing through movement (Figure 4.4b).
   - *Through and right turn (RMT):* A right-turning vehicle merging on the same lane as a through vehicle (Figure 4.4c).

(**a**) P2V: Conflict Type 1

(**b**) P2V: Conflict Type 3

(**c**) P2V: Conflict Type 5

(**d**) P2V: Conflict Types 2,4, and 6

Figure 4.3 P2V conflict types

- *U-turn and a Following Left turn (UFL):* A leading U-turn with a following left-turning vehicle (Figure 4.4d).
- *Right turn and a Following Through (RFT):* A leading right-turning vehicle with a following through vehicle (Figure 4.4e).
- *Lane Change and Adjacent Through (LCC):* A lane-changing vehicle conflicting with adjacent through (Figure 4.4f).
- *Rear-end Conflicts:* A leading vehicle moves slower than the following vehicle in the same lane.
- *A U-turn and an adjacent right turn.*

If one or more vehicles do not strictly follow traffic rules (e.g., run the red light), other conflicts are also possible, namely, adjacent through movements, left turn, and adjacent through. Some conflict types may be inherently more dangerous than other types. For example, the left turn and opposing through conflicts may lead to a more serious crash than a merging, a diverging, or a rear-end conflict. Further, the left turn and opposing through conflict is dangerous when the slow left-turning vehicle is the first to cross the conflict point. The less dangerous and common occurrence is when the left-turning vehicles yield to the through vehicles before completing the turn.

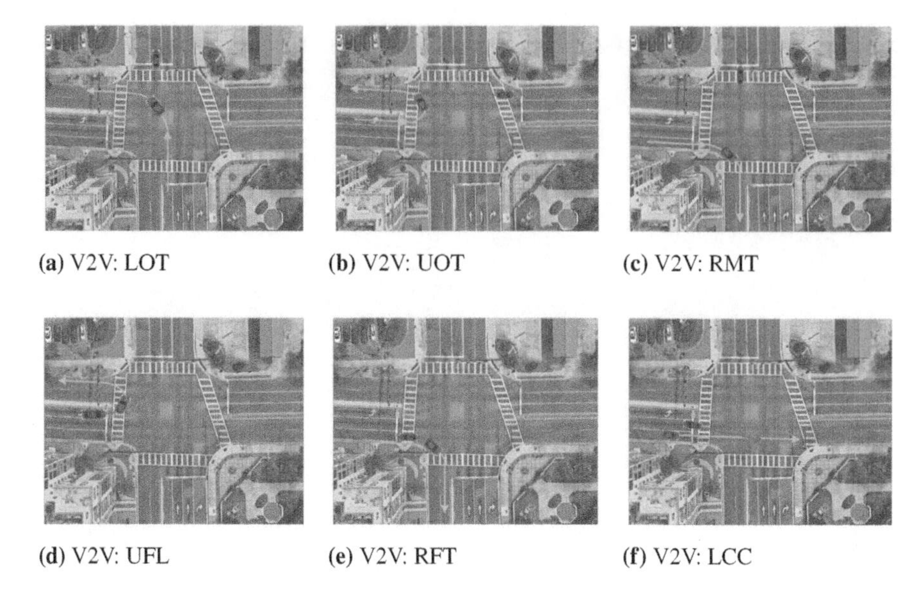
(a) V2V: LOT (b) V2V: UOT (c) V2V: RMT

(d) V2V: UFL (e) V2V: RFT (f) V2V: LCC

Figure 4.4 V2V conflict types

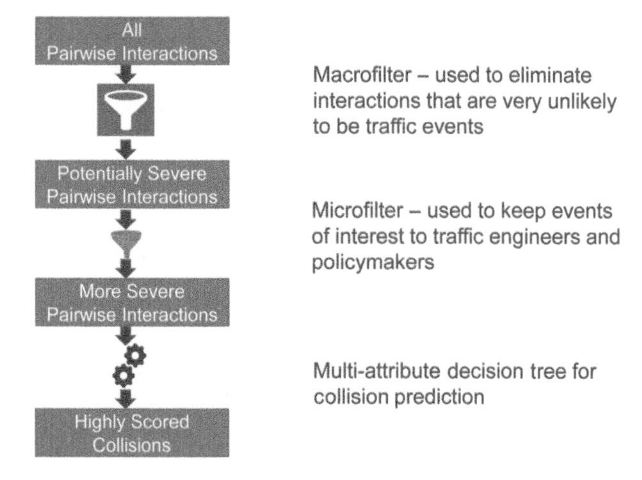

Figure 4.5 Event filtering sieve. The macrofilter checks for movement phases of conflicting trajectories, timing, and distance features. The microfilter checks finer aspects, such as whether yields were properly given

4.3.3 EVENT FILTERING

Figure 4.5 shows the multistage event filter we employ to prune the set of events. The macrofiltering stage checks the following conditions for two road users at the intersection in the same time frame: (1) Are they in a conflicting traffic phase? (2) Is the TTC or PET within a user-defined threshold, say, 10 seconds? (3) Are they spatially within or close to the intersection and within a user-defined threshold, say, 10 meters from each other? (4) Are both road users moving? The event passes through to the microfilter when all these conditions are satisfied. The first check for vehicle–vehicle interactions in the microfilter is whether an event is recorded more than once from separate TTC and PET computations. If the event did not result in a post-encroachment and there is no corresponding PET, the event must last for more than 1 decisecond to be considered severe. The second check for vehicle–vehicle interactions is whether each vehicle properly yielded to the other as per the traffic rules in case of a conflicting movement. If so, the event is filtered away.

For pedestrian–vehicle interactions, the first check in the microfilter is to find if the pedestrian is violating the pedestrian signal. If so, the event is considered severe, even if no vehicles are nearby. Highlighting the behavior allows practitioners to be aware of a pattern and adjust signal timing if needed. Suppose the pedestrian follows the signal, yet the event already made it through the macrofilter (indicating a conflicting maneuver); in that case, the microfilter checks if the distance between the pedestrian and the vehicle is 5 meters or less. If so, then the event is considered severe. On the other hand, if the pedestrian is about to enter the crosswalk and the vehicle is close by, the filter checks if the pedestrian's distance is less than 1 meter, and in that case, the event is considered severe.

All the thresholds mentioned in this section are easily configurable by the user based on specific intersection geometry and user characteristics.

4.3.4 EVENT MODELING

Although the event filters were efficient in pruning the event set, manually reviewing the filtered videos still constitutes a significant investment of human resources. For this reason, we utilize a simple algorithm to determine high-intensity regions in the feature space for severe events, where these events have been annotated manually. Because the event count after applying the filters is small compared to the total number of features for the events, there is a high chance of overfitting if a feature space with all the features is considered. So, to determine the high-intensity regions, we use only three features: speed, acceleration, and TTC or PET, as appropriate for that event. The algorithm plots 2D scatter plots of the events for each pair of features from the three-feature set and then sweeps the 2D space with a straight line to find the best intercept for which the line separates the severe and the non-severe events. The straight line acts as a separator of the 2D feature space, and we consider several such lines with different slopes to arrive at a near-optimal partitioning. This process is repeated for each partition if a partition is not purely from one class of events and contains a good mix of severe and non-severe events. In any case, the process is repeated up to a maximum of three levels of recursion. This algorithm could lead to more than one high-intensity region for the same dataset, which, once identified, can be conveniently used as a classifier for severe events. An example is presented in the experiments section. Human ratings of events are used to validate the model's predictions.

Table 4.1
Description of the Six Intersections Used for Our Analysis

ID	Intersection	City	Speed Limit (Major/Minor) Mph	Pedestrian Presence (% of Total Traffic)	Total Volume	Left-turn Type
1	University Ave and 13th St	Gainesville	35/25	18.5	146,133	Protected
2	University Ave and 17th St	Gainesville	25/25	41.8	67,550	Protected/Permissive
3	University Ave and 20th Dr	Gainesville	25/25	2.9	105,590	Protected/Permissive
4	NW 23rd Ave and NW 55th St	Gainesville	45/30	1.6	97,173	Protected/Permissive
5	Post Office and Rhinehart	Orlando	45/45	1.2	61,530	Protected/Permissive
6	Lake Mary and Rhinehart	Orlando	45/45	0.2	55,889	Protected

4.4 EXPERIMENTS

We applied our video processing algorithms end-to-end on six different intersections for the first week in November 2021. We collected data between 6 a.m. and 7 p.m. for each intersection, yielding 546 hours of video data, which were processed using our software and analyzed for intersection safety. Table 4.1 gives the intersection details. Based on video analysis, Table 4.1 also presents the total traffic and the percentage of pedestrians versus drivers. Three intersections are on an arterial adjacent to a university (ID: 1, 2, 3), one intersection is adjacent to a high school (ID: 4), and the two other intersections are in a city (ID: 5, 6). Though all these intersections have right-hand-drive traffic, our algorithms can also analyze intersections with left-hand-drive traffic.

We present our results on conflict analysis separately for pedestrians and vehicles in Sections IVA and IVB, respectively. While Table 4.1 gives an aggregate volume of the total number of pedestrians observed during the study period, we can further disaggregate the pedestrian count by the pedestrian phases and by day of the week and hour of the day. We present this analysis for the intersection of University Avenue and 13th Street, but the other intersections may be analyzed similarly. Four pedestrian phases, P2, P4, P6, and P8, and eight vehicle phases, 1–8, for the University Avenue and 13th Street intersection are shown in Figure 4.6.

Figure 4.7 shows the pedestrian volumes for the four pedestrian phases at the University Avenue and 13th Street intersection by day of week and hour of the day.

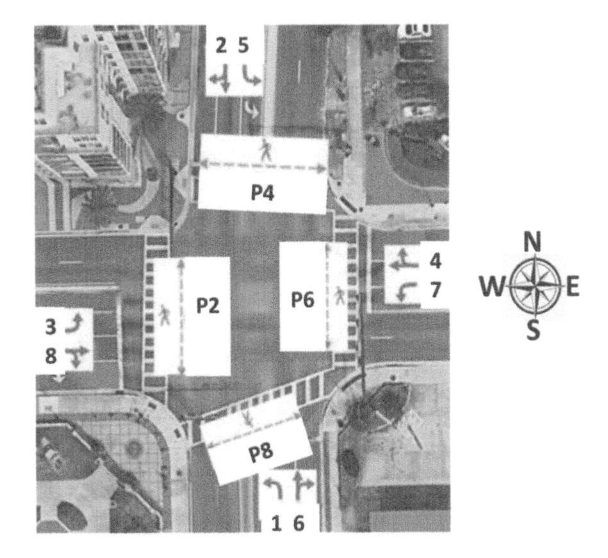

Figure 4.6 Four pedestrian phases, P2, P4, P6, and P8, and the eight vehicle phases, 1, 2, 3, 4, 5, 6, 7, and 8, are shown for University Avenue and 13th Street

We observe that (i) the volume in phase 4 is the highest, (ii) the volume on a Saturday, November 13, 2021, is the highest because there was a football game at the university stadium on University Avenue, (iii) the volumes on Monday and Tuesday are higher than the rest of the weekdays, and (iv) this intersection is large, so the pedestrians on the crosswalks closest to the camera are the best processed by the video processor. The crosswalks may be ordered as phases 4, 2, 6, and 8 by their proximity to the camera.

4.4.1 PEDESTRIAN–VEHICLE CONFLICT ANALYSIS

Figure 4.8 counts pedestrian–vehicle conflicts by the conflict type and cycle on the University Avenue and 13th Street intersection. The conflicts are shown by the day of the week and the hour of the day. We observe: (i) Conflict Type 1, which is a right-turning vehicle with a pedestrian on the adjacent parallel crosswalk, occurs most frequently; (ii) Conflict Type 1 happens throughout the day but is more likely to happen around 12 noon to 2 p.m.; (iii) Conflict Type 3, which is a left-turning vehicle with a pedestrian on the parallel opposite crosswalk, happens most frequently on game day; (iv) Conflict Type 5, which is a through vehicle with a pedestrian on the far crosswalk, a dangerous conflict, happens more frequently during 1–4 p.m. The vehicular movements can further analyze each conflict type, and Figure 4.9 shows a sample of such an analysis. We observe that the afternoons and weekends are when pedestrians are more prone to violate the traffic light and undertake dangerous crossings. For Conflict Type 1, we found that the westbound right (WBR) and southbound right (SBR) happen most frequently on weekdays and weekends. Such detailed

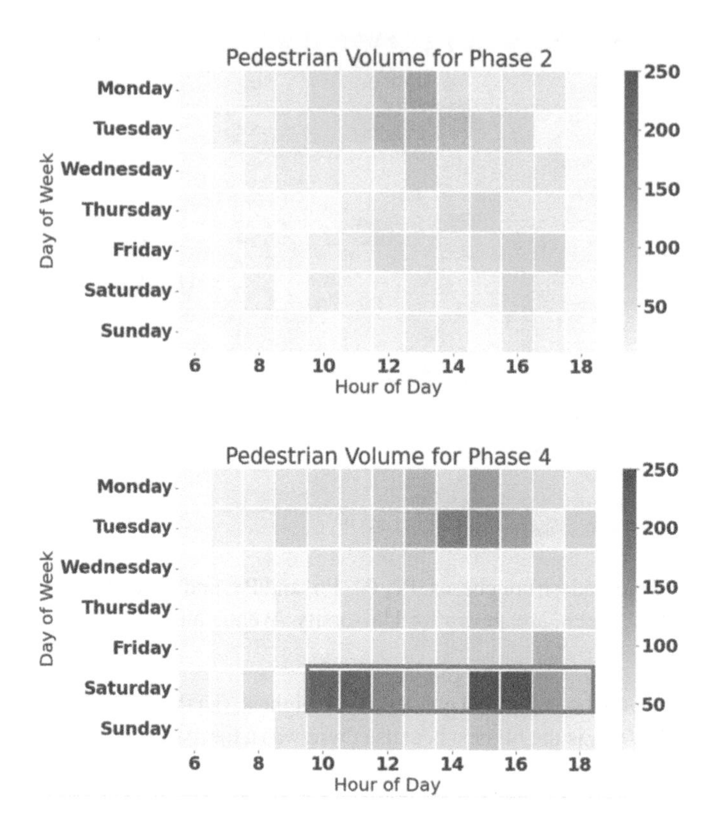

Figure 4.7 Pedestrian volumes by the four pedestrian phases at the University Avenue and 13th Street intersection. Among the weekdays, Monday and Tuesday are busier at this intersection. On Saturday, November 13, 2021, pedestrian traffic was especially high because of a football game being held at the University Stadium

analysis of pedestrian–vehicle conflicts could give insights into countermeasures to reduce conflicts.

4.4.2 VEHICLE–VEHICLE CONFLICT ANALYSIS

Table 4.2 shows the number of potentially conflicting interactions over a week and the performance of the two-level macro- and microfilters in filtering the events. The events include both vehicle–vehicle and pedestrian–vehicle conflicts. All pedestrian–vehicle events that remain after applying the microfilter are considered severe by default because of the vulnerable nature of pedestrians.

We manually verified for vehicle–vehicle events that 30–60% of these might be regarded as severe. We further applied our multi-attribute decision tree algorithm for classifying an unseen event automatically as severe or non-severe. The results of this step are presented in the Vehicle–Vehicle Multi-Attribute Decision Tree (V2V MADT) column that further reduces the count of severe events, which may be quickly

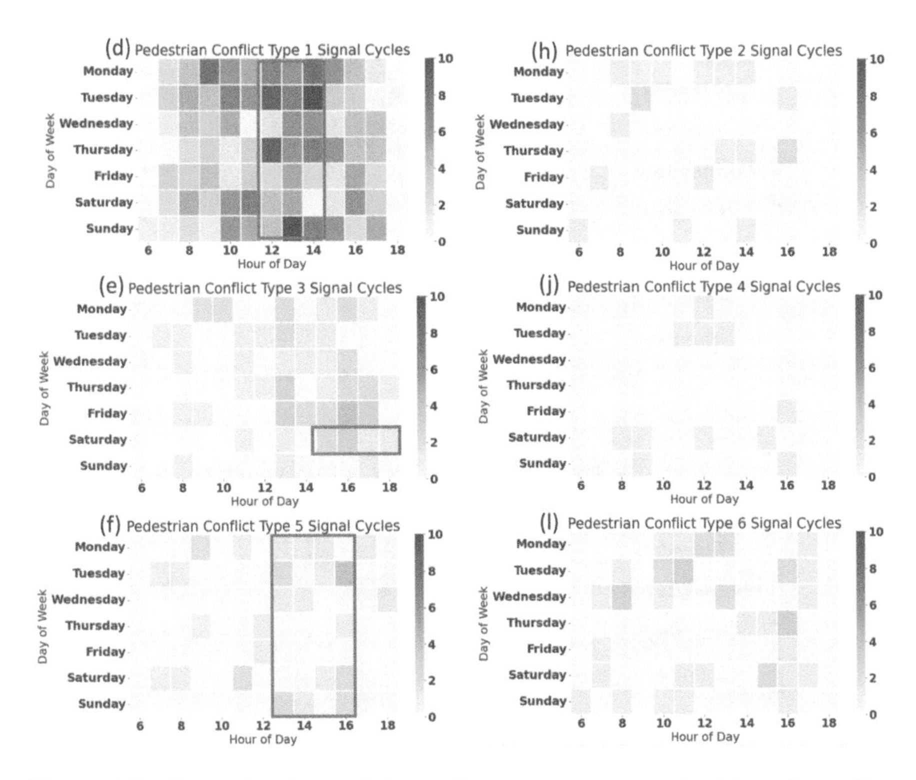

Figure 4.8 The pedestrian–vehicle conflict events are categorized into six conflict types and counted by day of the week and hour of the day. Figure parts (a), (b), and (c) give the raw event count for Conflict Types 1, 3, and 5, while parts (d), (e), and (f) give a count of signal cycles with at least one conflict. Conflict Type 1 occurs most frequently and is more likely between 12 noon and 2 p.m. Saturday, November 13, 2021, was a game day, and there was an uptick in the conflict counts from part (c). However, these conflicts are clustered, and fewer signal cycles are affected, as can be seen from part (d), Conflict Type 5, which is through vehicle with pedestrians in the far crosswalk, is a dangerous conflict that happens more frequently during 1–4 p.m.

evaluated manually for insights into applicable countermeasures. Thus, starting from millions of potential conflict interactions, our filtering scheme reduced the events to a small set of severe events.

Table 4.3 categorizes the vehicle–vehicle conflict types into the previously defined classes. Some conflicts did not belong to any of our defined conflict types. Among the known conflict types, left opposing through is the most severe as it could lead to significant damage to life and property if any of the conflicts results in a collision. The other conflict types are merging, or diverging conflicts, and any resulting collision yields a very low-impact crash. So, event categorization helps emphasize only the more dangerous conflict types by placing more weight on them. For example,

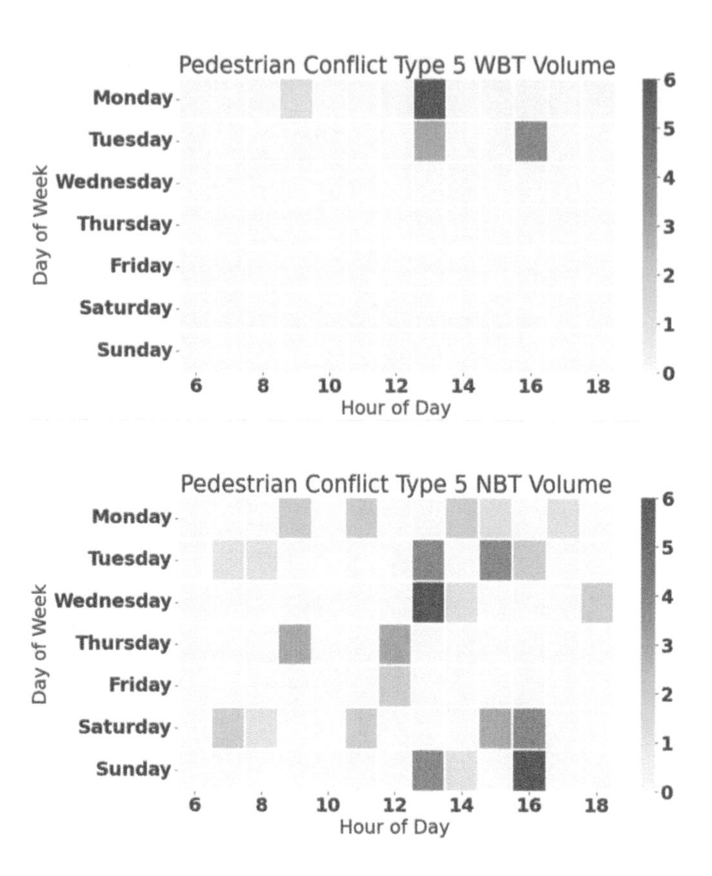

Figure 4.9 Volume of pedestrian–vehicle Conflict Type 5 by vehicle movement. This conflict type is the most dangerous between a through vehicle and a pedestrian in the far crosswalk. We have watched the video in almost all these cases and found that pedestrians violate their signals

in Table 4.3, the column titled "Total" is a simple sum of all conflict types, whereas the column titled "Weighted" gives a weighted total, where the left opposing through conflicts have been assigned a weight of 4. The "Normalized" column computes the conflict volume per 10,000 road users. These numbers are obtained by dividing the weighted conflicts by the corresponding exposure metric (explained in the next paragraph) and multiplying by 10,000. The "Normalized" conflicts allow us to rank the intersections by safety. For example, we can conclude from Table 4.3 that the Lake Mary and Rhinehart intersection is the safest. Table 4.4 shows the exposure metrics for the known conflict types. The exposure metric was computed as the sum of the number of vehicles participating in either of the two movements involved in a conflict. For example, if there is a conflict between northbound left (NBL) and southbound through (SBT), then the exposure metric for left opposing through will

Table 4.2

Total Number of Potentially Conflicting Interactions over a Week from Different Intersections and Performance of the Two-level Macro- and Microfilters in Filtering the Events to a Handful for Further Manual Analysis

Intersection	Conflicts	Macrofilter	Microfilter	P2V Events	V2V	V2V MADT
University Ave and 13th St	1,918,822	5,152	888	722	166	125
University Ave and 17th St	459,995	5,370	1,045	959	86	42
University Ave and 20th Dr	2,247,395	5,433	403	259	144	85
NW 23rd Ave and NW 55th St	947,921	5,938	217	63	154	112
Post Office and Rhinehart	362,354	344	95	67	28	28
Lake Mary and Rhinehart	1,279,019	956	28	0	28	28

After applying our multi-attribute decision tree algorithm, the column V2V MADT contains the event counts.

Table 4.3

Vehicle-to-vehicle Conflicts: (a) Total Sum; (b) Weighted Sum (Relative Weight of 4 Assigned to Left Opposing Through); (c) Normalized Based on Exposure (Per 10,000 Vehicles)

Intersection	Left Opposing Through	Right Merging Through	Right Following Through	Rear-End Conflict	Others	Total	Weighted	Normalized
University Ave and 13th St	15	2	37	28	84	166	211	24
University Ave and 17th St	33	2	15	9	27	86	185	21
University Ave and 20th Dr	91	5	9	3	36	144	417	62
NW 23rd Ave and NW 55th St	73	6	22	6	47	154	373	43
Post Office and Rhinehart	12	0	7	1	8	28	64	15
Lake Mary and Rhinehart	1	1	0	2	24	28	31	14

have a component that is the sum of all vehicles making NBL and SBT maneuvers. The exposure metric serves as a denominator in normalizing the conflict volume. Figure 4.10 shows the steps in our algorithm for isolating the high-intensity regions in the three-dimensional space (maxSpeed, maxDeceleration, minTime). For a conflict, maxSpeed defines the maximum speed of the two vehicles involved, while maxDeceleration is the maximum brake applied by either of the two vehicles. minTime is the minimum time for TTC or PET, which comes from the point representation of the involved vehicles and a bounding box representation of these vehicles. We take the minimum of these two times. The data used in Figure 4.10 are from the NW 23rd Avenue and NW 55th Street intersection. The steps in the algorithm are demonstrated in

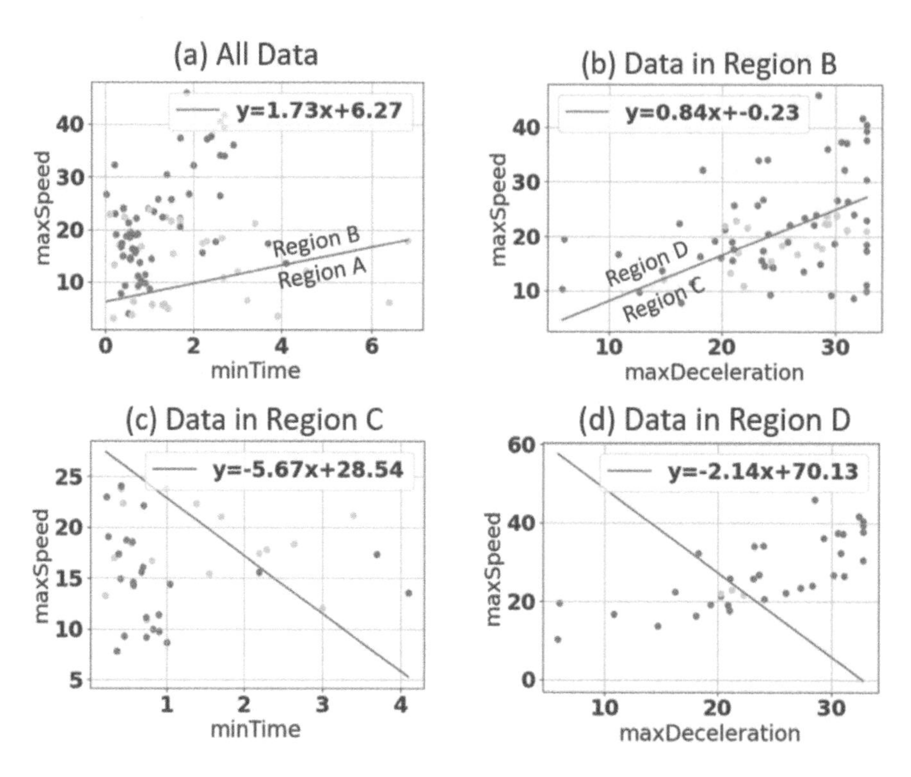

Figure 4.10 The steps in finding high-intensity regions of severe events. (a) All data and the line that separates them. In this diagram, the red dots represent severe events, while the blue dots represent non-severe events. The data here are from the NW 23rd Avenue and NW 55th Street intersection. Region A consists of non-severe points. (b) Region B is further split into regions C and D. Splitting regions C and D again, in parts (c) and (d), respectively, gives us high-intensity regions

Figure 4.10(a–d). The high-intensity region is obtained from randomly processing 80 of 100 filtered events from the intersections. The remaining 20 events are used to test the classifier. We manually annotated the 100 filtered events as severe or non-severe. The accuracy of this scheme is 90%, with a 92% recall and 90% precision. We used our algorithm separately for each intersection. The splits shown in Figure 4.10 are different for each intersection because the intersections have other characteristics such as speed limits, a school nearby, etc. The column V2V MADT in Table 4.2 shows the volume of events that were further pruned by our algorithm to find high-intensity regions. We did not apply the MADT algorithm to P2V events because pedestrians are vulnerable users, and all conflicts that remain after using the filters are considered severe by default. The accuracy of the MADT algorithm depends on the accuracy of video processing algorithms because video processing plays a crucial role in the computation of the safety features such as speed and deceleration.

Table 4.4
Traffic Exposure for Each Type of Conflicting Movement

Intersection	Left Opposing Through	Right Merging Through	Right Following Through	Rear-end Conflict	Total Vehicles
University Ave and 13th St	53,484	27,663	72,735	191,788	146,133
University Ave and 17th St	99,974	28,030	63,826	179,082	67,550
University Ave and 20th Dr	66,774	59,706	31,596	121,966	105,590
NW 23rd Ave and NW 55th St	97,810	45,988	36,814	63,782	97,173
Post Office and Rhinehart	47,759	23,568	20,459	23,294	61,530
Lake Mary and Rhinehart	8,450	8,884	*N/A*	11,542	55,889

4.5 DISCUSSION

This chapter developed a systematic and novel methodology for analyzing intersection safety based on video analysis. We developed algorithms that use video analysis and signal timing data to perform accurate event detection and categorization in terms of the phase and type of conflict for both pedestrian–vehicle and vehicle–vehicle interactions. We introduced a new surrogate safety measure, severe event, quantified by multiple metrics such as TTC or PET as recorded in the event, deceleration, and speed. We developed an efficient multistage event filtering approach followed by a multi-attribute decision tree approach that prunes the extensive set of conflicting interactions to a robust set of severe events.

Using our analysis and based on the limited number of intersections, we found that the dominant conflicts at intersections with heavy pedestrian use are right-turning vehicles on the adjacent parallel crosswalk. We could identify the specific right-turn directions contributing to this problem and the hours during the week when the problem peaks. Categorization of vehicle–vehicle interactions showed that for intersections with permissive left turns, the more common conflict is between a left-turning vehicle and an opposing through vehicle. The intersections with protected left-only displayed some merging and diverging conflicts, inherently less severe.

5 Performance–Safety Trade-offs

5.1 INTRODUCTION

Intersection safety studies are essential because traffic intersections are prone to crashes leading to injury or fatality [87]. According to USDOT, more than 50% of road crashes leading to fatality or injury happen at or near traffic intersections. With rapid technological advancement and price drops [88], video cameras are now commonly deployed as sensors in traffic intersections. Existing intersection safety assessment methodologies often require the analysis of historical data to infer current and future intersection user behavior. Although helpful, these data are often biased toward what has been reported, which is retrospective and often incomplete. Video analysis of traffic intersections has no reporting or survivorship biases because the analysis times from the fisheye camera [89] are precise and unbiased. This chapter presents a methodology to systematically use the trajectory data from video analysis, severe events data, and signal phasing data to detect and analyze conflict hotspots and evaluate the efficacy of countermeasures to improve intersection safety. The trajectory and severe events data are available from our previous work [90], which implemented video processing and severe event detection algorithms. The signal phasing data are available from the advanced traffic controllers (ATC) at traffic intersections.

We present a novel evaluation engine. The engine ingests trajectory data, severe events data, and signal phasing data at a given intersection for different days of the week and times of the day.

The evaluation engine has three primary data processing modules: The first module computes pedestrian and vehicle volume hotspots; the second module computes pedestrian–vehicle (P2V) and vehicle–vehicle (V2V) conflict hotspots; the third evaluates intersection service as a proxy to intersection performance for the study period. Multiple signal phasing configurations may be systematically analyzed using the evaluation engine to develop volume and conflict heatmaps and a performance-safety trade-off chart for the intersection. Traffic engineers may use this trade-off chart to select a configuration that optimizes intersection safety and performance.

More detail is presented below about the three primary data processing modules:

1. The volume hotspot detection module computes the peaks and troughs in pedestrian and vehicle volume. It is helpful to understand where the volume peaks and spikes are and if there are valleys and troughs in the volume. This information helps a traffic engineer implement signal timing changes to address safety issues only during the peak period instead of throughout the day.

DOI: 10.1201/9781003431176-5

2. The conflict hotspot detection module computes the temporal hotspots of P2V and V2V conflicts by the conflict types and movements of the involved trajectories. P2V conflicts occur when pedestrians and vehicles come dangerously close, and their two trajectories intersect. V2V conflicts are defined as the occurrence of evasive vehicular actions and are recognizable by braking or weaving maneuvers [91]. If a hotspot pattern emerges in the conflict analysis, a traffic engineer could apply a countermeasure to address those conflicts. The evaluation engine uses a novel visualization scheme to simultaneously represent temporal conflict hotspots with the spatial locations within the intersection where the conflicts occur. We use the reciprocal of the number of conflicts as a measure of intersection safety.
3. The intersection-service evaluation module does a fine-grained aggregation of vehicle volume at subcycle levels and outputs an intersection service histogram that counts the number of vehicles entering an intersection in 5-second bins since the start of green. Vehicle volume is collected for 60 seconds. Vehicles from all cycles within the study period are aggregated to arrive at a single histogram. We use intersection-service evaluation as a proxy for a measure of intersection performance [92]. Specifically, we use the entering vehicle counts for all movements between the 5- and 15-second marks to represent the performance of the intersection. The rationale for aggregating the vehicle counts for 10 seconds, choosing the start and end points as 5- and 15-second marks, respectively, is that we exclude the start-up loss and measure the volume before saturation headway. This fine-grained aggregation of vehicle volume captures the effect of any temporal issues impacting intersection performance that may otherwise be unnoticeable in hourly aggregations. Examples of temporal issues impacting the performance may be the presence of many pedestrians at the intersection or a change in signal phasing configuration. We use the performance measure to compare two signal phasing configurations at the same intersection.

The evaluation engine with the modules above may be used for studying performance–safety trade-offs from multiple signal phasing configuration scenarios. We can analyze the current scenario using the evaluation engine and arrive at an appropriate countermeasure to potentially improve safety or performance issues at the intersection. We demonstrate the operation of the evaluation engine on two different intersections using more than a week of data on each intersection. One of these intersections is near a high school with teenage pedestrians and drivers. The other intersection has a pedestrian presence of more than 40% of the total users.

The key contributions of this chapter are described as follows:

1. We have developed systematic end-to-end software to analyze intersection data to find intersection safety and performance metrics.
2. We have formalized the simultaneous treatment of intersection safety and intersection performance, producing performance versus safety trade-off charts.

3. Additionally, our graphical heatmap output is beneficial to figure out the temporal hotspots for pedestrian–vehicle and vehicle–vehicle conflicts and the spatial locations. The existing literature does not describe this concise representation that simultaneously captures the conflicts' temporal and spatial properties.

The rest of the chapter is organized as follows. Section 5.2 presents the related work, while Section 5.3 presents the background for work. Next, Section 5.4 presents our system's built-in methodology and how we can use it systematically to discover and address safety issues. Section 5.5 illustrates how to apply the steps to find the safety issues for both pedestrian–vehicle (P2V) and vehicle–vehicle (V2V) interactions at two different intersections. Finally, we discuss the key findings in Section 5.6.

5.2 RELATED WORK

Traffic intersections are more at risk for near-miss events and accidents. This section first describes surrogate safety measures: What and how they are beneficial. The sensors available today to compute the surrogate safety measures are discussed next, and, finally, the existing work on traffic safety analysis using video processing is presented.

5.2.1 SURROGATE SAFETY MEASURES

Surrogate safety measures are indicators that strongly correlate to traffic conflicts [93, 94]. Surrogate safety measures effectively detect near-miss events, which occur much more frequently than crashes and are generally not reported through traditional channels. Crash analysis studies require years of data because of the infrequent nature of actual traffic collisions [95, 96]. Combined with the fact that crash data are often incomplete due to the under reporting of crashes and injuries and that the geometry of the intersection may change over such long periods, the effectiveness of crash analysis studies diminishes significantly. Surrogate safety studies, on the other hand, can uncover important safety issues using about a week of video footage, making it a time - and cost-effective solution for monitoring traffic intersections.

The surrogate safety measures most commonly used for near-miss detection are time-to-collision (TTC) and post-encroachment-time (PET). TTC is the time remaining to prevent a collision (by applying brakes, steering away, or using some other preventive action), measured precisely as the difference between the time the road user takes action to the point where the collision can occur [74]. PET is the difference between when the first road user leaves a point and the second road user reaches that point [75]. Lower values of TTC and PET indicate higher risks of collision [97]. After computation of TTC and PET, one way to identify severe events is by applying thresholds on TTC and PET [85, 81]. Several other proposed measures are based on either spatial proximity or acceleration–deceleration patterns of vehicles [76, 77, 78]. Our software considers TTC and PET (thresholded values) with other measures such as speed, acceleration, and distance between the two road users when a severe event happens. The thresholds we use are 2 seconds for TTC and 3 seconds for PET. These values are configurable so that the user can set them to different ones. The early definitions of traffic conflicts indicate that only the most extreme of traffic interactions have been considered safety-critical [98]. Hyden [79] proposed a hierarchy of traffic events varying in severity. Arun et al. [78] note that there is still no consensus on what constitutes a safety-critical event or a near-miss. Vogel [99] compares headway and TTC as safety indicators, showing that these two measures are independent. Peesapati et al. [100] evaluated and found that PET is effective as a surrogate measure for a left turn and opposing through conflicts. Feng et al. [101] found a strong correlation between factors leading to near-misses and those leading to crashes, which highlights the benefits of analyzing the surrogate safety measures for severe near-miss events.

Johnsson et al. [102] find surrogate safety measures appropriate for vulnerable road users, including pedestrians and bicyclists. Fu et al. [103] developed a framework for assessing the safety of pedestrian–vehicle interactions. Chen et al. [104] developed lane-based models for evaluating pedestrian–vehicle interactions at nonsignalized crosswalks.

After fusing surrogate safety measures with crash analysis data, Yang et al. [105] created a new surrogate measure, risk status. Surrogate safety measures are also used extensively in traffic simulation models [93, 106] and connected autonomous vehicles [107, 108].

5.2.2 INTERSECTION SENSORS

Loop detectors [109] are traditionally installed at intersections for monitoring traffic and signaling states. While these detectors are very effective in detecting traffic incidents [110], it is not possible to get the exact location of the conflict from these under-the-ground detectors. With recent technological advancements, various sensors, such as video, lidar, and infrared (IR) cameras, are being used for intersection monitoring [12, 111, 112].

Video-based traffic monitoring is increasingly popular because it is relatively low-cost and fast to process and analyze [113]. Unlike data from loop detectors, video footage gives us rich information about the object trajectories, object classification (e.g., car, bus, truck, motorcycle, pedestrian), precise location and severity of traffic conflicts, and traffic violations. Unlike lidar and IR cameras, video cameras are easier to deploy and maintain [114]. Video data are also easier to review and understand than lidar and IR camera outputs. On the downside, video cameras depend on proper lighting conditions at intersections. The two main types of video cameras are ordinary cameras and fisheye cameras. The advantage of using fisheye cameras is that a single camera can monitor the whole intersection for smaller intersections, and two cameras are sufficient for the slightly bigger ones. On the other hand, with ordinary cameras, a traffic intersection would require as many cameras as there are approaches (typically four) to monitor the intersection completely [115].

The viability of video cameras for traffic monitoring has encouraged the development of robust video processing algorithms and surrogate safety measures. Hence, intersection safety assessment using video analysis is an active research topic. We will briefly review the existing work in this area. We reuse the video processing framework developed in our previous work [90]. Section 5.3 provides the details of our video processing algorithm.

The differences between this work and our previous work [90] are as follows. The software [90] was used for processing the videos and generating a database of trajectory data and severe events over multiple days of the week and times of the day. In this chapter, we develop an evaluation engine that uses the trajectory, severe events, and signal phasing data to perform a spatiotemporal analysis over multiple intersections to identify temporal conflict hotspots and volume hotspots. This chapter presents a visual representation of P2V and V2V conflicts that annotates heatmaps with symbols that give the practitioner clues about the spatial locations of the conflicts and the movements of the involved trajectories for further analysis. Additionally, the evaluation engine determines overall performance and safety metrics for a given signal phasing configuration scenario and plots a performance–safety trade-off chart for an intersection for multiple scenarios. Such a chart would be helpful to a traffic engineer for the selection of a countermeasure because it would give an idea of the performance–safety characteristics of that or similar configurations.

5.2.3 INTERSECTION SAFETY ANALYSIS USING VIDEO CAMERAS

This section presents a collection of existing work on traffic intersection monitoring using video cameras [116]. Ultimately, we compare the current work to two other recent papers on safety assessment using video analysis.

Saunier et al. [117] developed a framework to estimate collision probabilities and their spatial distributions based on video data collected in Kentucky. Ismail [118] treats the use of computer vision techniques to process traffic video data for intersection safety analysis. Stipancic et al. [119] develop and evaluate the crash frequency and severity models, incorporating GPS-derived surrogate safety measures as predictive variables. St-Aubin [120] developed a thesis studying computer vision techniques for traffic roundabouts.

Our work is closely related to the work by Samara et al. [85], where the authors use video analysis for analyzing intersection traffic. The main difference is that the work [85] has not treated P2V conflicts. Further, we can identify trajectories that help to pinpoint any persistent issues with specific movements. Moreover, the work [85] uses only PET as a safety indicator and does not discuss phase-based conflict hotspots for conflicts. Neither is there a treatment for countermeasures to improve intersection safety nor how intersection safety impacts intersection performance.

Kronprasert et al. [121] study safety performance using a video-based traffic conflict analysis system. They do not use fisheye cameras but rather regular ones that monitor the direction they point to, which requires multiple cameras. Moreover, the videos from all the cameras must be merged to get complete information about the intersection trajectories. Furthermore, the surrogate safety measure used [121] was TTC only. Our work presented in this chapter considers P2V conflicts. For P2V and V2V conflict hotspots, we do a more advanced level of conflict categorization regarding the involved trajectories.

5.3 BACKGROUND

Given a video captured by a fisheye camera at an intersection, there are six steps in our video analytics system – process the footage to extract trajectories that are timestamped (x, y) coordinates; fuse the trajectories with signaling information; process the trajectories to mine features that help with safety analysis; find all P2V and V2V conflict events; filter the events to retain only the "severe events"; and categorize the event based on the involved trajectories. These steps have been presented previously [90] and are included here for completeness.

5.3.1 VIDEO ANALYSIS

With a fisheye video as input, the video analysis software draws a bounding box around each object in a video frame, identifies the object's class based on several previously annotated images, and finds the (x, y) coordinates of the object as the center point of the bounding box. A video frame is a static snapshot. The fisheye cameras installed at the intersections capture 10 video frames per second. The object detection and tracking module utilizes YOLOv4 [122] to detect different road participants, including cars, buses, trucks, pedestrians, and motorcyclists. A modified DeepSORT algorithm [52] associates detections across frames and assigns a unique ID for each object. The (x, y) coordinates are in the circular fisheye space. A fisheye lens has a wide angle, creating a panoramic or hemispherical nonrectilinear image. So we apply a post-processing step to map the coordinates to rectilinear space using fisheye-to-perspective transformation followed by thin-plate spline (TPS) warping [123, 124, 68]. The new timestamped coordinates in the rectilinear space are stored in a database (DB) table.

5.3.2 HIGH-RESOLUTION CONTROLLER LOG ANALYSIS

The high-resolution controller logs [125] provided by the appropriate agency are analyzed to extract signal phasing information for an intersection. Fusing the signal phasing with object trajectory coordinates over the time axes gives us a holistic view of a trajectory based on when the object arrived and left the intersection. We use this information later to count the vehicles that pass the stop bar in any one direction in intervals of 5 seconds after the traffic light turns green.

5.3.3 FEATURE COMPUTATION

The following is a list of the key features we compute:

1. *Standard Near-miss Attributes:* We compute the standard risk assessment metrics for every event, such as TTC and PET.
2. *Signal Phase Information:* The fused video and signal phasing dataset is used to determine features such as the current vehicle signal, current pedestrian signal, and if the event occurs during the beginning (first 10% of the cycle), middle, or end (last 10% of the cycle) of the current signaling phase.

3. *Trajectory features:* The trajectory-related features are the direction of motion, associated phases, and lanes or crosswalks.
4. *Speed Features:* These include the current speeds and accelerations for vehicle–vehicle interactions.
5. *Distance:* Spatial distance between two users at the time of the conflict.

5.3.4 CATEGORIZATION OF SEVERE EVENTS

The categorization of the P2V and V2V conflicts is described in this section and used in the rest of the chapter.

1. *P2V Conflicts:* The following are the main conflicts between vehicles and pedestrians at signalized intersections:
   - *Conflict Types 1 and 2:* Right-turning vehicle with the pedestrian in an adjacent parallel crosswalk (Figure 4.3a) and near-side crosswalk (Figure 4.3d), respectively.
   - *Conflict Types 3 and 4:* Left-turning vehicle with the pedestrian in the far-side (parallel opposite) crosswalk (Figure 4.3b) and near-side crosswalk (Figure 4.3d), respectively.
   - *Conflict Types 5 and 6:* Through vehicle with a pedestrian in the far-side crosswalk (Figure 4.3c) and the near-side crosswalk (Figure 4.3d), respectively.

   Among these conflicts, Conflict Types 1 and 3 are possible conflicts between pedestrians and vehicles at signalized intersections if all vehicles and pedestrians strictly follow the traffic rules (assuming a three- or four-leg intersection, which are the most common).
2. *V2V Conflicts:* As the primary purpose of signalization is to reduce or eliminate conflicting movements at the intersection, the following are the possible conflicts between vehicles at signalized intersections if all vehicles strictly follow the traffic rules (assuming a three- or four-leg intersection, which are the most common):
   - *Left turn and opposing through (LOT):* A left-turning vehicle in a permitted phase conflicts with an opposing through movement (Figure 4.4a).
   - *U-turn and opposing through (UOT):* A U-turning vehicle in the permitted phase conflicts with an opposing through movement (Figure 4.4b).
   - *Merging right and through (RMT):* A right-turning vehicle merging on the same lane as a through vehicle (Figure 4.4c).
   - *U-turn and a following left turn (UFL):* A leading U-turn with a following left-turning vehicle (Figure 4.4d).
   - *Right turn and a following through (RFT):* A leading right-turning vehicle with a following through vehicle (Figure 4.4e).
   - *Lane change conflict with adjacent through (LCC):* A lane-changing vehicle conflicting with adjacent through (Figure 4.4f).
   - *Rear-end conflicts (REC):* A leading vehicle moves slower than the following vehicle in the same lane.
   - *A U-turn and an adjacent right turn (URT).*

If one or more vehicles do not strictly follow traffic rules (e.g., run the red light), other conflicts are possible, namely, adjacent through movements, left turn, and adjacent through. Some conflict types may be inherently more dangerous than the other types. For example, the left turn and opposing through conflicts may lead to a more serious crash than a merging, a diverging, or a rear-end conflict. Further, the left turn and opposing through conflicts are more dangerous when the slow left-turning vehicle is the first to cross the conflict point. The less dangerous, relatively common occurrence is when the left-turning vehicles yield to the through vehicles before completing the turn.

5.4 METHODOLOGY

Using trajectory data and severe events from multiple videos over different hours of the day and days of the week, we compute temporal volume and find conflict hotspots. The spatiotemporal analysis helps discover safety issues in the intersection and when and where they occur. This knowledge is crucial for arriving at effective countermeasures with minimal impact on the rest of the traffic.

5.4.1 EVALUATION ENGINE MODULES

Figure 5.1 shows our evaluation engine and the various engine modules with their inputs and outputs. The evaluation engine modules are described in detail in the following sections.

5.4.1.1 Volume Hotspot Detection Module

The volume hotspot detection module is the first module of the evaluation engine that estimates (1) pedestrian volume on the different crosswalks and (2) volume peaks and troughs by the time of day and day of the week. For example, this module will help answer a query such as "How many pedestrians crossed using the north crosswalk leg of intersection X between 12 p.m. and 1 p.m.?" or "When does pedestrian volume aggregated on an hourly basis exceed a threshold T on intersection X for a given period Y?" Thus, the pedestrian hotspot detection module exposes the times of day when the pedestrian volume peaks and provides a guideline to a traffic engineer about the time of day when countermeasures for alleviating potential P2V conflicts may be implemented. Usually, such countermeasures penalize intersection performance, so implementing such countermeasures in targeted time intervals instead of throughout the day may benefit pedestrians and vehicles.

Similarly, the vehicle hotspot detection module may be used to compute vehicle volume. In contrast to video cameras, the loop detectors installed in the intersection cannot distinguish between through and right-turning vehicles. The vehicle hotspot detection module exposes times of the day and days of the week when the vehicle volume peaks for a given movement. The volume hotspot detection module could output turning movement counts (TMC) for all movements over user-selected time intervals. For example, the module can output the total number of vehicles with

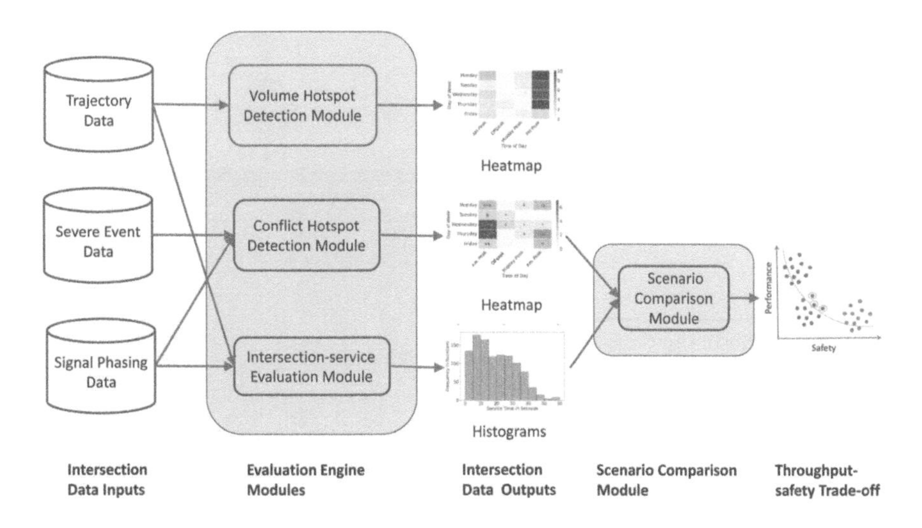

Figure 5.1 Evaluation engine modules with their inputs and outputs

movement NBT, NBL, NBR, SBT, SBL, SBR, EBT, EBL, EBR, WBT, WBL, and WBR, every 15 minutes during the AM peak.

5.4.1.2 Conflict Hotspot Detection Module

The second module in the evaluation engine computes the temporal hotspots of P2V and V2V conflicts and the conflict types. Examples of conflict types are, for P2V, a left-turning vehicle with a pedestrian on the parallel opposite crosswalk (Figure 5.2b), and for V2V conflicts, a left-turning vehicle conflicting with an opposing through vehicle (LOT, Figure 5.3b). Section 5.3.4 lists conflict types in the "severe event" database.

This chapter develops a visualization scheme to describe the spatial location where conflicts happen in the intersection. This is illustrated in Figures 5.2 and 5.3. Given the conflict type and the quadrant information, it is possible to derive the movements of the trajectories involved in the conflicts. For example, for the P2V conflict chart in Figure 5.2a, there are a few conflicts on the east crosswalk during Monday's AM peak. Since the type of the P2V conflict is a vehicle turning left with a pedestrian on the parallel opposite crosswalk, we can derive the vehicle movement as SBL, as illustrated in Figure 5.2b.

Similarly, the V2V conflict chart in Figure 5.3a shows a LOT conflict in the SW quadrant on Monday's midday peak. This implies that the movements of the conflicting trajectories are WBL and EBT, as demonstrated in Figure 5.3b. If a specific combination of phases emerges repetitively from this analysis, the traffic engineer could selectively address those. Otherwise, a countermeasure that impacts all phases must be applied if all phases have similar conflicts. Discovering such patterns is crucial for arriving at effective countermeasures.

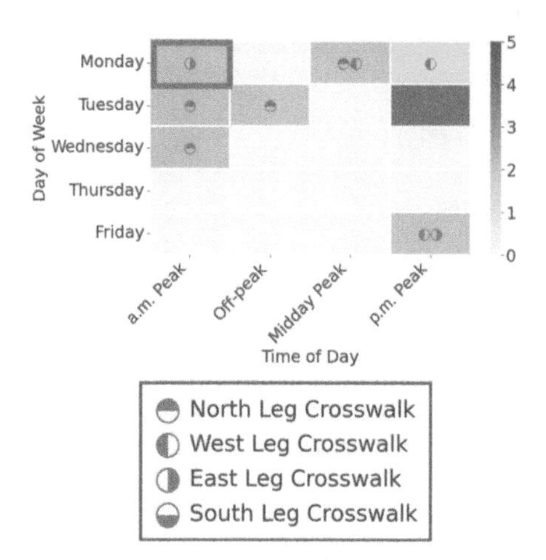

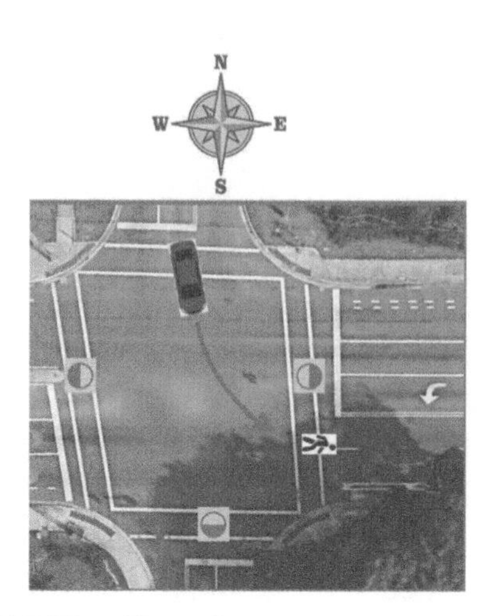

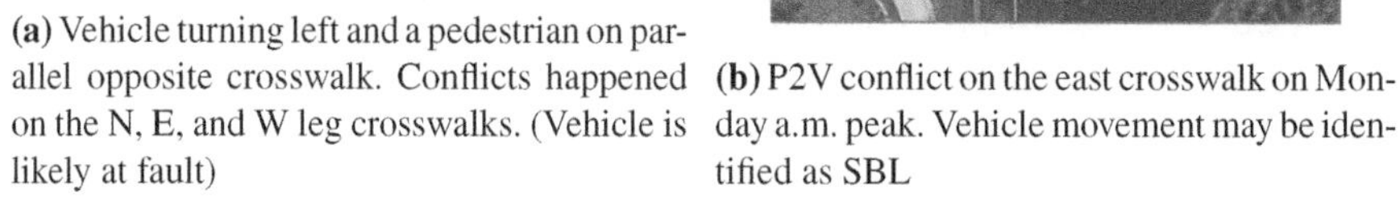
(a) Vehicle turning left and a pedestrian on parallel opposite crosswalk. Conflicts happened on the N, E, and W leg crosswalks. (Vehicle is likely at fault)

(b) P2V conflict on the east crosswalk on Monday a.m. peak. Vehicle movement may be identified as SBL

Figure 5.2 The spatial location of P2V conflicts is overlaid on the heatmap. Given the P2V conflict type, one can derive the trajectories from the overhead view of the crosswalk (b)

The intersection safety measure is defined as the reciprocal of the sum of conflicts over the study period. The sum may be weighted to put more emphasis on the severe conflicts. For example, a sample weighting scheme could assign a weight of 10 to P2V conflicts of any type, a weight of 5 for V2V LOT conflicts, and a weight of 1 for all other conflict types.

5.4.1.3 Intersection Performance Evaluation Module

The evaluation engine assesses intersection performance by analyzing the vehicle volume in 5-second intervals starting from the beginning of the green light, creating service histograms. Alternatively, it can use 10-minute aggregations of vehicle trajectories. This module can generate histograms and heatmaps for specific lanes, phases, or combinations using trajectory data instead of loop detector data. This enables analysis of all lanes, regardless of whether a loop detector is installed.

5.4.1.4 Scenario Comparison Module

Using data generated by the modules described so far for multiple scenarios, the evaluation engine plots a point for each signal phasing configuration scenario on a performance–safety graph. For example, applying a countermeasure would change performance and safety, generating a new point. The horizontal axis of the

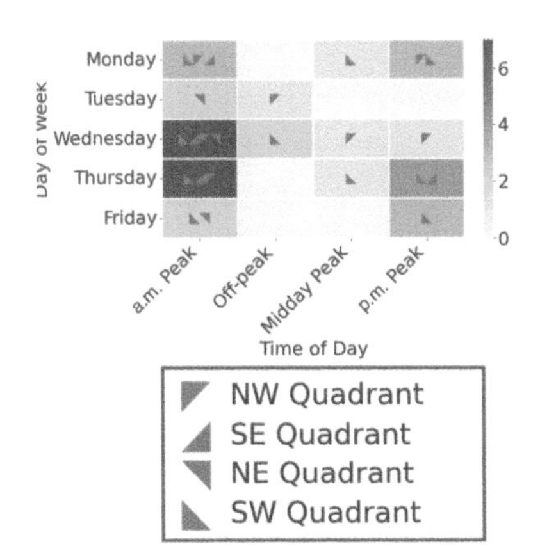

(a) V2V LOT conflict. Conflict happens on all quadrants during Wednesday a.m. peak. LOT conflicts are usually more dangerous

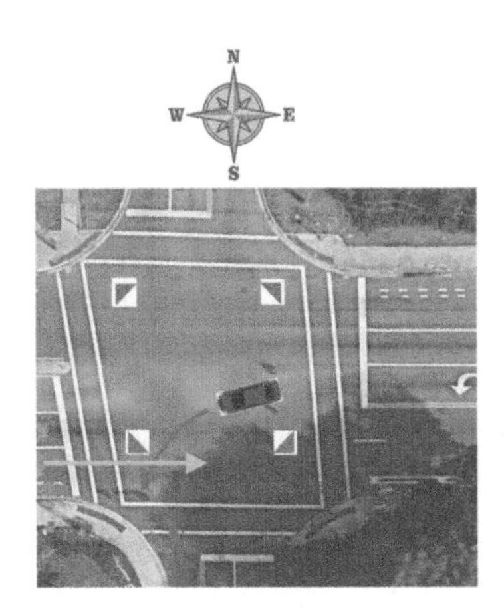

(b) V2V Conflict on SW quadrant during Monday midday peak. Vehicle movements may be derived as WBL and EBT

Figure 5.3 The spatial location of V2V conflicts is overlaid on the heatmap. Given the V2V conflict type, once can derive the movements of the trajectories from the quadrant involved

performance–safety graph represents intersection safety, while the vertical axis represents intersection performance. The performance–safety graph created from the multiple scenarios under study may be used to select a good set of signal phasing configurations for further consideration.

Figure 5.4 shows an example of a performance–safety graph. A rough sketch of what this might entail is as follows: If an intersection has a high performance and no severe events, it is ideal. It is acceptable if an intersection has low performance with no severe events. It is also potentially acceptable with high performance and a few severe events (still avoiding crashes). However, it is unacceptable if an intersection has a low performance and multiple severe events. Thus, using the Pareto curve in the performance–safety graph, a traffic engineer can select a configuration that improves intersection safety without a substantial drop in intersection performance.

Some possible countermeasures are as follows:

1. Change in signal phasing or sequencing pattern [126].
2. Implementation of leading pedestrian interval (LPI), which typically gives pedestrians a 3- to 7-second head start when entering an intersection with a corresponding green signal in the same direction of travel [127].
3. Implementation of exclusive pedestrian phasing (EPP), which stops all vehicular movement and allows pedestrians to cross in any direction at the intersection [127].

We demonstrate the use of the evaluation engine for improving intersection safety in Section 5.5.

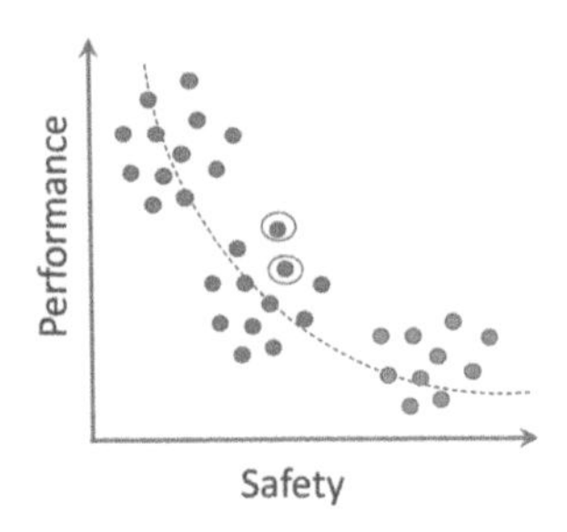

Figure 5.4 Performance–safety trade-off. Each dot represents a scenario. The optimal scenarios are circled

Table 5.1
The Details of the Intersection Used in the Case Study

ID	Intersection	Speed Limit (mph)	Pedestrian Presence (%)	Left Turn Type	Flashing Yellow Arrow	Right Turn on Red
1	University Ave. and 17th St.	25/25	41.8	Protected/ Permissive	Yes	Yes

The intersection is located in Gainesville, FL. The column "Speed Limit" gives the major/minor street speed limits.

5.5 EXPERIMENTS

This section presents a case study on a representative traffic intersection under different pedestrian and vehicle demand conditions. This case study demonstrates our evaluation engine's effectiveness in analyzing an intersection's safety and performance and the impact of any countermeasures in improving the safety of an intersection.

Characteristics of the intersection are shown in Table 5.1. Videos from this intersection were collected for a week in November 2021, between 6 a.m. and 11 p.m. Thus, over 100 hours of video footage were collected and processed for the spatiotemporal study. We collected several videos at other times for the evaluation of countermeasures. The high-resolution controller logs were also collected for information about signal phasing during the same period. The speed limit column gives the major/minor street speed limits, while the pedestrian presence column in Table 5.1 provides the percentage presence of pedestrians among all users, pedestrians, and vehicles.

The observations and analyses for the intersection are presented here based on a time-of-day segmentation obtained using the changes in the signal timing patterns. The time segments are defined as follows:

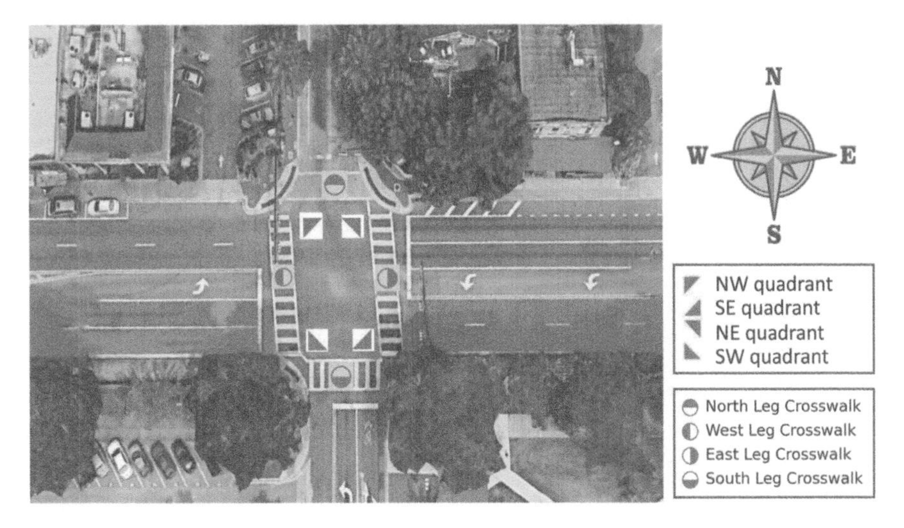

Figure 5.5 Intersection 1: Crosswalks and intersection quadrants are marked. The University of Florida (UF) is at the south of the intersection. University Avenue, the arterial adjacent to UF, runs east to west. The intersection has a heavy pedestrian and vehicle presence. The size of the intersection is quite small, and it serves many pedestrians throughout the day and night and a large number of vehicles. The main issue at this intersection is P2V conflicts

For weekdays:	
a.m. peak	07:00–09:30
Off-peak	09:30–11:00
Midday peak	11:00–14:30
p.m. peak	14:30–18:00

This chapter will refer to intersections by ID, and because we are presenting one representative intersection, we will use ID 1. We begin with a detailed image of the intersection, study the pedestrian volume during weekdays, identify potential problem areas and solutions, and evaluate feasible countermeasures. Detailed charts for the intersection representing vehicle volume are omitted for conciseness. Instead, the peak vehicle volume in any one direction is cited as a reference point.

5.5.1 INTERSECTION 1

Figure 5.5 shows Intersection 1 near the UF. This intersection has a maximum volume of more than 700 vehicles per hour (EBT, p.m. peak). We start by analyzing the volume and conflict characteristics at this intersection.

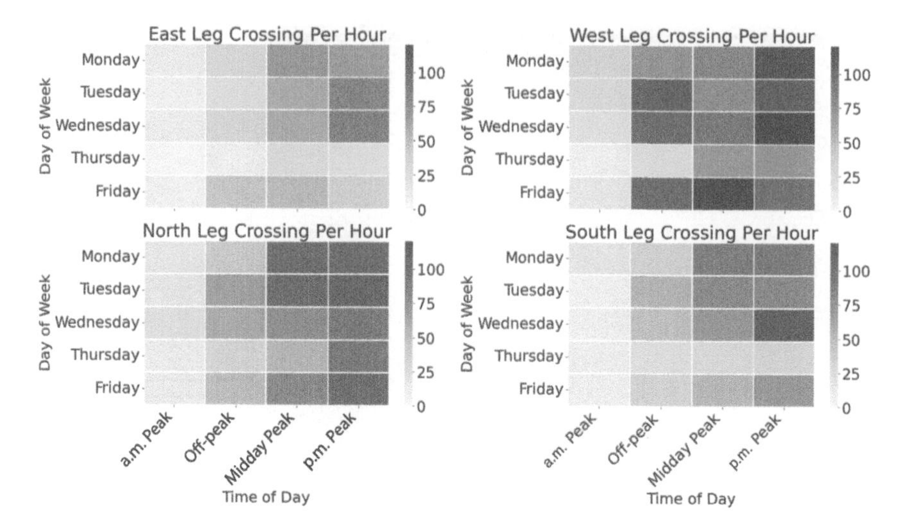

Figure 5.6 Intersection 1: Pedestrian volume during the weekdays. There is an increase during midday and p.m. peaks, coinciding with lunch hour and class dismissal times. At its peak, there are about 120 pedestrians per hour. The pedestrian volume is higher on the west crosswalk because more restaurants are accessible across the west leg. The gradient color scale on the right displays the number of vehicles

5.5.1.1 Pedestrian Volume

Figure 5.6 shows the pedestrian volume during the weekdays. There is increased pedestrian traffic during the midday and p.m. peaks, coinciding with lunch hour and class dismissal times. Thursday has sparse traffic in the morning for the Veteran's Day holiday, though the volume picks up later in the day. At its peak, there are about 120 pedestrians per hour using crosswalks. The pedestrian volume is higher on the west crosswalk because more restaurants are accessible across the west leg.

5.5.1.2 P2V conflicts

Figure 5.7 shows P2V conflicts during the weekdays. Many conflicts exist on all crosswalk legs, particularly involving left-turning vehicles with pedestrians on the parallel opposite crosswalk, as shown in 5.7a. This intersection has protected/permissive left turns. The vehicles often initiate a left turn in the permissive phase, even in the presence of pedestrians on the parallel opposite crosswalk. Pedestrians sometimes violate their walk signal and start crossing while a protected left turn is being served.

There were several other conflicts where a right-turning vehicle did not yield to a pedestrian on the adjacent parallel or the near crosswalk (Figure 5.7c). The conflicts involving a through vehicle and a pedestrian (Figure 5.7d) are likely the pedestrian's fault because vehicles tend to follow their signals at this intersection.

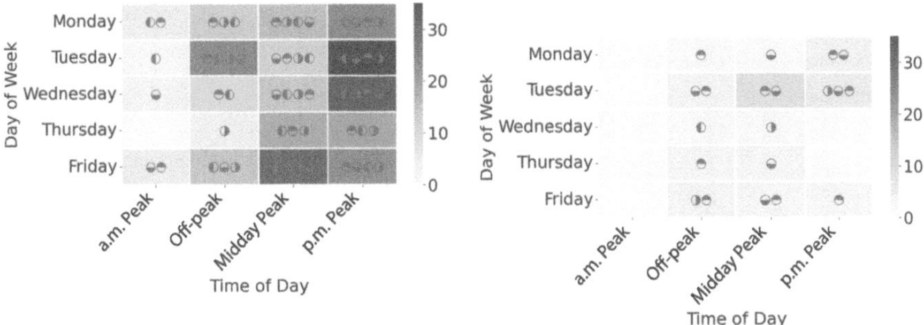

(**a**) Vehicle turning left and pedestrian on parallel opposite crosswalk. (Vehicle likely at fault)

(**b**) Vehicle turning left and pedestrian on near crosswalk. (Pedestrian likely at fault)

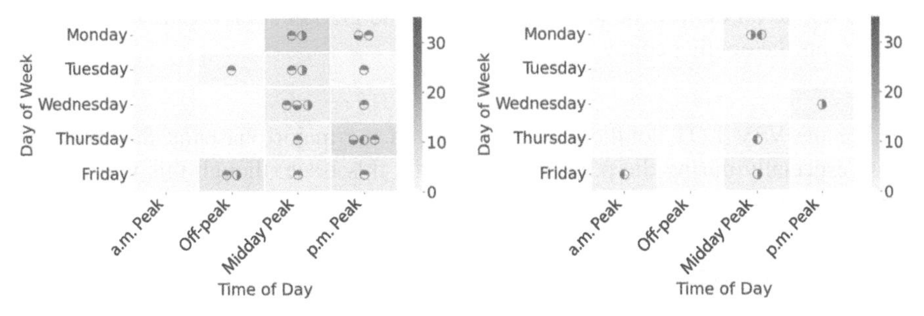

(**c**) Vehicle turning right and pedestrians on near crosswalk. (Vehicle likely at fault)

(**d**) Through vehicle and pedestrians on far crosswalk. (Pedestrian likely at fault)

Figure 5.7 Intersection 1: P2V conflicts during the weekdays. Overall, there are many P2V conflicts at this intersection. The gradient color scale on the right displays the conflict count

5.5.1.3 V2V conflicts

Figure 5.8 shows the V2V conflicts during the weekdays. There are far fewer V2V conflicts compared to P2V conflicts. We can see that the NE quadrant is affected during the a.m. peak for RFT conflicts because there is a heavy inflow of people going to work. Both the NE and the SW quadrants are high in LOT conflicts during the p.m. peak because many cars take EBL and WBL permissive turns, and the uptick in WBT and EBT traffic during the p.m. peak compounds the problem.

5.5.1.4 Suggested Countermeasures

The excessively high P2V conflicts at this intersection may be addressed by applying LPI or EPP in the afternoon to address the conflicts in Figure 5.7a and b. A no-right-turn-on-red directive may reduce the conflicts involving right-turning vehicles in Figure 5.7c.

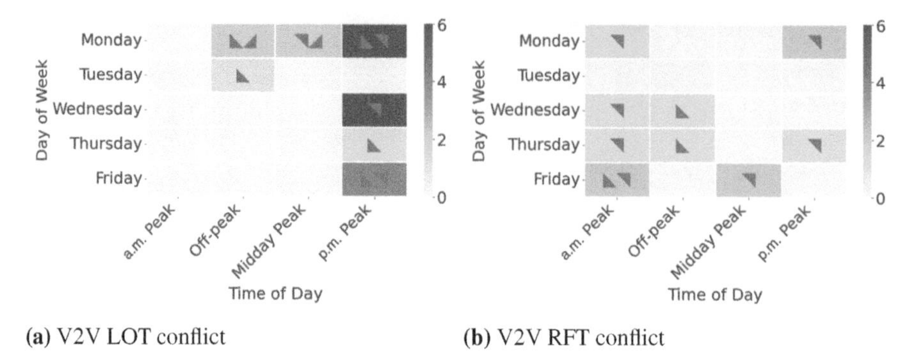

(a) V2V LOT conflict (b) V2V RFT conflict

Figure 5.8 Intersection 1: V2V conflicts during the weekdays. There are far fewer V2V conflicts compared to P2V conflicts. The p.m. peak is particularly affected, most likely due to an outflow of workers from UF. The gradient color scale on the right represents the conflict count

For the V2V LOT conflicts, split phasing of the major phases at this intersection, especially during the p.m. peak, might fix this issue, though this will impact performance.

5.5.1.5 Countermeasure Evaluation: EPP

EPP is implemented in this intersection on Thursday, Friday, and Saturday evenings, 8 p.m. to 1 a.m., because the P2V conflicts may be very dangerous at night. Specifically, many bars and nightclubs in the area draw the predominantly young college student population. The evening traffic for 1 hour, 8–9 p.m. was collected and analyzed for Wednesdays (no countermeasure) and Thursdays (with EPP countermeasure). The P2V conflicts are presented in Figure 5.9a and b, respectively. The benefit of the EPP is clearly seen: Most P2V conflicts are resolved in Figure 5.9b. The only conflict (on Thursday, April 21, 2022) happened because the pedestrian violated the walk signal. The pedestrian volume is illustrated in Figure 5.10a and b for Wednesdays and Thursdays, respectively.

Performance–Safety Trade-off: The performance is impacted while the EPP effectively resolves P2V conflicts. To evaluate the approximate impact on the performance of the intersection, we measured the volume of vehicles on both days and find fewer vehicles per 10-minute intervals with EPP implemented (Figure 5.10a and b). However, the decrease in volume was not particularly large. This indicates that the performance cost of the countermeasure is minimal. Therefore, the added safety of the exclusive pedestrian phase is well worth the slight inconvenience to the vehicles. Table 5.2 aggregated the P2V conflicts and vehicle volume on Wednesdays when there were no countermeasures to Thursdays when the EPP countermeasure was in effect.

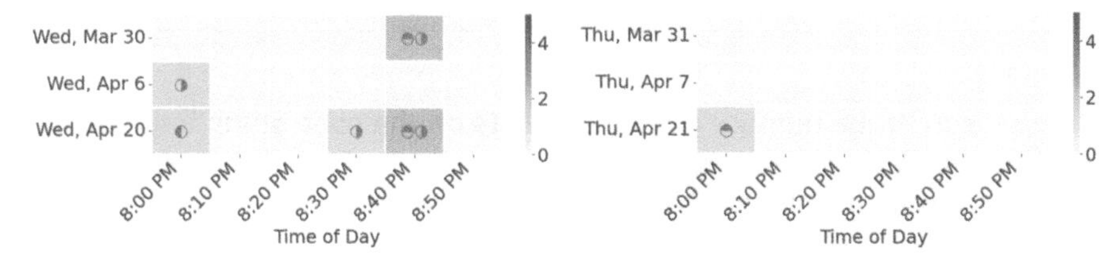

(a) P2V conflicts on three Wednesdays without countermeasure

(b) P2V conflicts on three Thursdays with countermeasures implemented

Figure 5.9 Intersection 1: Countermeasure EPP. From 8 p.m. onward, EPP is implemented on Thursday, Friday, and Saturday. The heatmap on the left shows the P2V conflict distribution on Wednesdays and compares that with the countermeasure implemented on Thursdays. The reduction in the number of conflicts is pronounced. The one conflict on April 21 was due to a pedestrian violating signals. The gradient color scale represents the conflict count

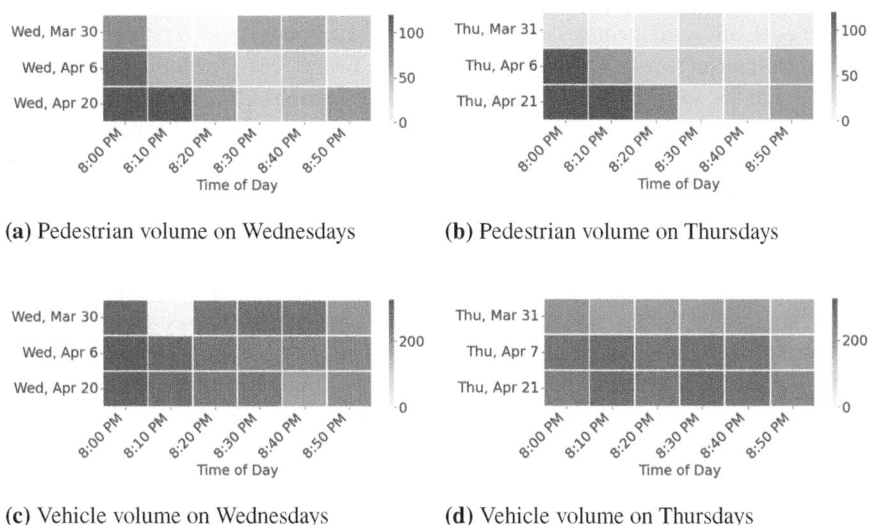

(a) Pedestrian volume on Wednesdays

(b) Pedestrian volume on Thursdays

(c) Vehicle volume on Wednesdays

(d) Vehicle volume on Thursdays

Figure 5.10 Intersection 1: Pedestrian and vehicle volume during the countermeasure study period. On Wednesday, March 30, between 8:10 p.m. and 8:20 p.m. the video collected was corrupted; hence, it was dropped from the analysis. The gradient color scale represents the vehicle count

5.6 DISCUSSION

Road safety is highly relevant in today's societies, as vehicle collisions claim lives every day. Intersections especially are prone to crashes which lead to the worst outcomes. Recent rapid improvements in video collection and processing technologies

Table 5.2

Comparison of the number of P2V conflicts on the case study intersection on Wednesday evening when EPP is not implemented and Thursday evening when EPP is implemented. Note the significant decrease in conflicts at a similar volume of traffic

Conflicts		Vehicle Volumes	
Wednesdays	Thursdays	Wednesdays	Thursdays
7	1	4,355	4,002

enable us to identify and alleviate dangerous situations. Conflict and volume hotspots are detected, and intersection performance is analyzed using trajectory data, severe events, and signal phasing. This chapter's extensive analysis and discussion of a representative intersection demonstrate the selection of appropriate countermeasures. Additionally, we evaluate the efficacy of countermeasures in an understandable and replicable manner.

Our evaluation engine takes inputs from the intersection – trajectory, severe event, and signal phasing data. The inputs are generated using our previous work on processing intersection videos [90] and from the intersection ATC. The evaluation engine has three primary modules: The first computes pedestrian and vehicle volume hotspots; the second calculates pedestrian–vehicle (P2V) and vehicle–vehicle (V2V) conflict hotspots; the third evaluates intersection service as a proxy to intersection performance for the study period. The evaluation engine then outputs histograms and conflict and volume hotspot charts, where each conflict chart is annotated with information that identifies the traffic phases involved in the conflicts. These outputs allow for possible use by researchers and traffic engineers. We also demonstrate how to compute performance–safety trade-off charts. The knowledge about performance–safety trade-offs is crucial for arriving at effective countermeasures with minimal impact on intersection performance. The readability of the trade-off graph makes it easy to compare and select appropriate countermeasures. We demonstrate the operation of the evaluation engine on a busy intersection – "Intersection 1" – adjacent to a university campus, using more than a week of data. More than 100 hours of video were collected and processed in the demonstration of our engine, and with the level of automation achieved, more videos can be analyzed at little cost.

Intersection 1 is located near UF and has a high pedestrian presence. The intersection has a maximum vehicle volume of more than 700 vehicles per hour during the p.m. peak. The number of pedestrians on the west leg is consistently the largest because of local restaurants, reflected in the more significant number of P2V conflicts with pedestrians on the west leg. The number of V2V conflicts is relatively small and is primarily associated with people coming to and going from work. A large number of P2V conflicts and the inherent danger due to evening conflicts influence the implementation of an EPP countermeasure. The countermeasure decreases the number of P2V conflicts a great deal. Furthermore, the impact on performance is insignificant relative to the safety enhancement at the intersection, justifying the continued use of the countermeasure.

This chapter aims to help the practitioner find any safety issues within an intersection that may not have resulted in a crash but could result in one in the future. This work results from a close interactive feedback loop between researchers and traffic practitioners. For example, our analysis uncovered the everyday pedestrian–vehicle conflict situations observed in Intersection 1, where the pedestrian was given a walk signal, and the left-turning vehicles were shown a permissive left a few seconds later. The conflicts did not lead to crashes or injuries but often to near-miss conflicts. This information finally allowed the practitioners to test the countermeasure of exclusive pedestrian signaling in Intersection 1. As a result of implementing the exclusive pedestrian phases, vehicles are stopped on all legs, and pedestrians may cross in any direction. This made the intersection safer for pedestrians.

A direction of potential future study is the automation of additional steps in the process we have outlined, including, but not limited to, the suggestion, analysis, and implementation of countermeasures on the fly. Also, the precept of a performance-safety trade-off in intersection analysis developed here can be applied further and potentially more broadly in road safety.

6 Trajectory Prediction

6.1 INTRODUCTION

Road intersections show complex geometry and traffic rules and require drivers' timely maneuvers. The Federal Highway Administration (FHWA) has reported that more than 50% of fatal and injury-causing crashes occur at or near intersections. Of all the intersection-related crashes, only about 4% are caused by vehicle or environmental reasons, and 96% are attributed to drivers [128]. Factors contributing to driver-attributed crashes at intersections include inadequate surveillance, false assumptions regarding other drivers' actions, obstructed views, illegal maneuvers, internal distractions, and misjudgments of gaps, speed, etc. [128]. Potential solutions for reducing crashes focus on providing driver assistance in controlling the vehicle or giving timely warnings of potential risks. To this end, predictive algorithms are proposed. For example, by tracking the surrounding vehicles, one can estimate their intended maneuvers, predict their future motion, and estimate the risk of collision. Previous studies [129, 130, 131, 132] have focused on the automated vehicle (AV) and advanced driver assistance system (ADAS) solutions. These approaches rely on onboard sensors like LiDAR, radar, and vision sensors. However, the onboard sensors have limitations, such as limited detection range or field of view, low resolution, drifting position estimation, and sensor imprecision. Therefore, we propose incorporating assistance from existing and planned intersection video infrastructure to improve safety. An essential basis of the information such systems will provide is real-time knowledge of the motions of other vehicles in the intersection and their potential to collide with other vehicles, i.e., trajectory prediction.

In this chapter, we tackle the problem of robust and flexible vehicle trajectory prediction at intersections from surveillance videos, aiming to give early warnings to vehicles about possible collisions and abnormal behaviors. Our proposed approach only requires vision sensors – no signal data are needed – and can be applied to signalized and unsignalized intersections. Our approach requires a setup stage, namely, (i) Google Maps alignment, (ii) learning typical motion patterns from historical data, and (iii) training the trajectory prediction model. After the setup, our model can make real-time predictions of all the vehicles' trajectories within the intersection. We also consider the noise the automatic vehicle tracking algorithms introduced and the varying lengths of trajectories captured.

The main contributions of the chapter are as follows:

- We present real-time vehicle trajectory prediction from surveillance cameras without requiring traffic signals or GPS data.
- Our algorithm automatically learns the typical motion patterns from historical data and representatives of them are used in the online phase to speed up and boost the performance of trajectory prediction.

DOI: 10.1201/9781003431176-6

- We design a deep-learning-based trajectory prediction model capable of dealing with variable-length observations and prediction periods.
- The model gives multiple reasonable predictions if multiple motion patterns are likely to describe the observed trajectory.

The rest of this chapter is organized as follows. Section 6.2 presents the related works on trajectory prediction methods. Section 6.3 gives an overview of our proposed approach, consisting of offline learning and online prediction phases. Section 6.4 describes the methodology of offline trajectory clustering and prototype generation. Section 6.5 presents the CCS transformations and the trajectory prediction model. Section 6.6 reports our experimental settings and quantitative and qualitative results. We conclude with a few observations in Section 6.7.

6.2 RELATED WORK

Vehicle motion or trajectory prediction has been studied extensively over the last two decades and is mainly used in ADAS and AV. The work in Lefèvre et al. [133] organized motion prediction approaches into three categories: physics-based, maneuver-based, and interaction-aware. Our approach falls into the maneuver-based category. The work in Ref. [133] further divides the maneuver-based approaches into prototype-based and intention estimation. Our approach can be classified as prototype based.

6.2.1 PROTOTYPE-BASED TRAJECTORY PREDICTION

Prototype-based approaches assume that vehicle motion can be grouped into a finite set of clusters. A prototype trajectory is learned for each cluster as its representative. In the case of four-way intersections, there are mainly 12 motion patterns, namely, *left turn*, *right turn*, and *going straight* from the four directions. For road sections, typical motion patterns are *lane keeping*, *lane change to the left*, and *lane change to the right*. An early work [134] adopts a statistical approach where the mean and standard deviation represent clusters of motion patterns. Later works [135, 136, 137] represent motion patterns with Gaussian processes (GP) due to their ability to capture spatiotemporal characteristics of traffic situations. One drawback of GP models lies in their computational complexity. As a nonparametric approach, the number of parameters grows as more training samples are provided. In a surveillance setting, a nonparametric approach will be slow and expensive to leverage the vast amount of data captured from cameras. In this work, our proposed approach fits a compact model to the training data and incorporates the prototype trajectories to make more informed predictions.

6.2.2 A RECURRENT NEURAL NETWORK (RNN) FOR TRAJECTORY PREDICTION

Trajectories are sequences of locations, velocities, etc. Recurrent neural networks (RNNs), especially long short-term memory (LSTM) and gated recurrent units

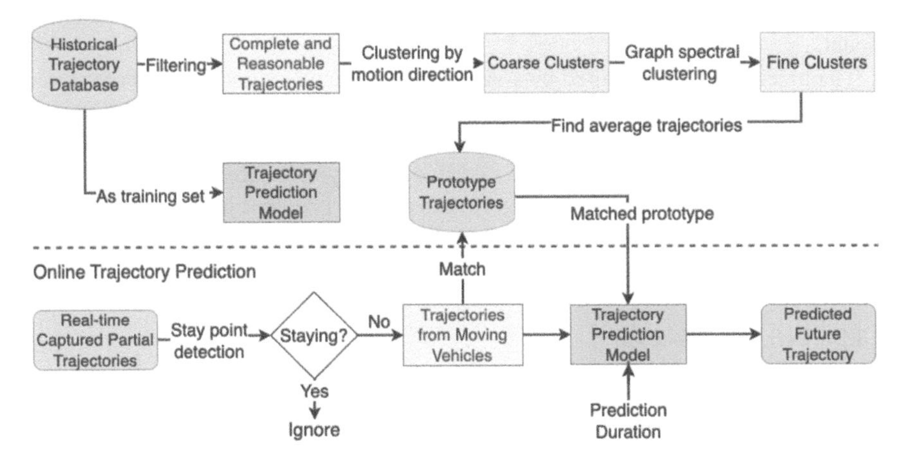

Figure 6.1 The overall workflow of trajectory prediction in offline and online phases

(GRUs), have been explored and broadly evaluated for the task of sequence generation and prediction. Over the last five years, RNN and its variants have been gaining popularity in trajectory prediction. The work in Alahi et al. [138] first applied an LSTM model to learn general human motions and predict their future trajectories. Later works [139, 140, 141, 142] extended the idea to vehicle trajectory predictions, enabling vehicles to reason about the future motion of other surrounding vehicles. To the best of our knowledge, no work utilizes a deep learning–based trajectory prediction model from surveillance vision systems. Unlike the above approaches for autonomous vehicles or ADAS, where modeling interaction is essential, our approach will be applied to give early warnings to vehicles entering intersections; hence, we focus on typical motion pattern finding and efficient online functioning.

6.3 SYSTEM OVERVIEW AND PIPELINE

The chapter is divided into two main components – an offline phase in which trajectories are clustered and promising prototypes are found. This is followed by an online step in which different LSTMs are constructed for trajectory prediction. The second phase necessarily uses the prototype information from the first phase. In this section, we give a procedural description of our algorithm workflow. Figure 6.1 concisely summarizes the workflow.

6.3.1 OFFLINE PHASE

The offline phase can be performed periodically to capture the dynamics of traffic evolution (e.g., road construction that causes vehicle detour) or done once for setup if the intersection geometry remains unchanged. There are two main tasks in the offline phase: (i) finding common motion patterns and their representatives and (ii) training the LSTM-ED (long short-term memory encoder–decoder) trajectory prediction model.

The trajectories are captured by a vision-based tracking algorithm and are stored in the database. We refer to the trajectories in the database as historical trajectories. We use a portion of historical trajectories to learn common motion patterns using a coarse-to-fine clustering algorithm and find prototypes to represent each motion cluster. A filtering and smoothing algorithm is applied before clustering to generate longer and smoother prototype trajectories. Lastly, the trajectory prediction model to be used in the online phase is trained in the offline phase.

6.3.2 ONLINE PHASE

The online phase receives real-time captured trajectories from the online tracking algorithm. We first detect whether the vehicle is waiting for traffic signals using the stay point detection algorithm from Cai et al. [143]. We start making predictions when the vehicle starts moving. We match a partial trajectory from a moving vehicle with prototype trajectories from the offline phase based on distance measurements. We then feed the partial trajectory, matching prototypes, and duration to be predicted into the trajectory prediction model to output its future trajectory.

6.4 TRAJECTORY CLUSTERING AND PROTOTYPE TRAJECTORIES

This section introduces the methods used to cluster trajectories and find prototypes.

6.4.1 HISTORICAL TRAJECTORY CLUSTERING

Vehicles at an intersection exhibit common motion patterns, and their trajectories can be grouped into clusters. One or more finer clusters can be found within each cluster of the same moving direction, as there might be multiple entering and exiting lanes of the same motion. Our clustering approach is, therefore, a two-step process where we first cluster the trajectories based on their direction of motion and next cluster the trajectories for a given motion using spectral clustering. In this section, we describe the clustering based on motion direction, followed by the description of a new distance measure for spectral clustering. Finally, we describe our spectral clustering approach.

6.4.1.1 Clustering by Motion Direction

When we analyze a collection of trajectories, one of the first steps is automatically identifying the trajectory phase. As shown in Figure 6.2, the vehicle phase system assigns a single number for both through and right-turn movements. We find it helpful first to have a separate category for the right-turn vehicular movement. So, in addition to the eight bins, each storing trajectories of a particular phase, we append four more bins for storing the trajectories making right-turns for phases 2, 4, 6, and 8.

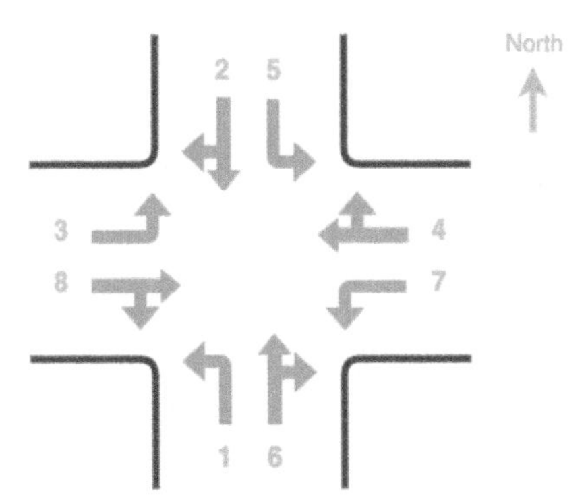

Figure 6.2 Phase diagram showing vehicles at four-way intersections as defined in Ref. [69]

The motion direction of a trajectory can be represented as the vector connecting the start and end points. This process was explained in detail in Chapter 3.

After partitioning the trajectories by motion direction into bins, we apply spectral clustering to the trajectories of each bin. We used a new distance measure for computing the distance between two trajectories in the same bin. This is described next.

6.4.1.2 Clustering by Graph Spectral Clustering

Spectral clustering is applied next to the trajectories in the same direction bins, and we get clusters of trajectories with the same movement. Spectral clustering may also be used to identify outliers or anomalous trajectories that are not similar to any other trajectories in the same bin. The details of our spectral clustering implementation may be found in Ref. [144].

6.4.2 PROTOTYPE TRAJECTORY GENERATION

From our clustering algorithm, clusters of motion patterns are found. This section presents our approach to generating prototype trajectories for each cluster.

6.4.2.1 Complete Trajectory Determination

Many trajectories extracted from the traffic video are incomplete, meaning they either appear from the middle of the intersection or disappear within it. These trajectories are as helpful as complete ones in the LSTM-ED model training. However, we prefer complete trajectories for prototype generation for two reasons: (1) complete trajectories are spatially aligned because they start and end in the same regions, and (2) the distance measure we use performs better with a longer reference trajectory.

Trajectories that start at an intersection entrance and end at an intersection exit are considered complete. We annotate the intersection boundary as a polygon. For

Figure 6.3 Complete trajectories in a cluster and its prototype

each motion pattern, we annotate its enter line and exit line. We consider a trajectory complete only when its starting point is close to the entering line, and its ending point is close to the exit line. The left part of Figure 6.3 shows the complete trajectories from one cluster.

6.4.2.2 Averaging Complete Trajectories

Given a set of complete trajectories from a cluster, we aim to find a representative for the cluster by averaging the trajectories. We represent the complete trajectories as 2D cubic splines, denoted as S, where x and y coordinates of a trajectory are parameterized by t:

$$\begin{aligned} S(t) &= (x(t), y(t)), \\ x(t) &= a_x t^3 + b_x t^2 + c_x t + d_x, \\ y(t) &= a_y t^3 + b_y t^2 + c_y t + d_y, \end{aligned} \tag{6.1}$$

where t's are equally spaced break points in the range of $[0, 1]$. As the complete trajectories are roughly spatially aligned, an average trajectory for each cluster can be found by

$$S_c(t) = \frac{1}{m_c} \sum_{j}^{m_c} S_j(t), \tag{6.2}$$

where S_j is a complete trajectory from the given set from cluster c, and m_c is the number of trajectories. After obtaining the average trajectories, the last step is to reparametrize and rescale the splines. We adopt the arc-length parameterization algorithm from Ref. [145], which results in equally spaced control points. We then rescale the splines so that consecutive points of a spline have a distance of 1 meter. Figure 6.4 shows the prototypes we obtained for an intersection.

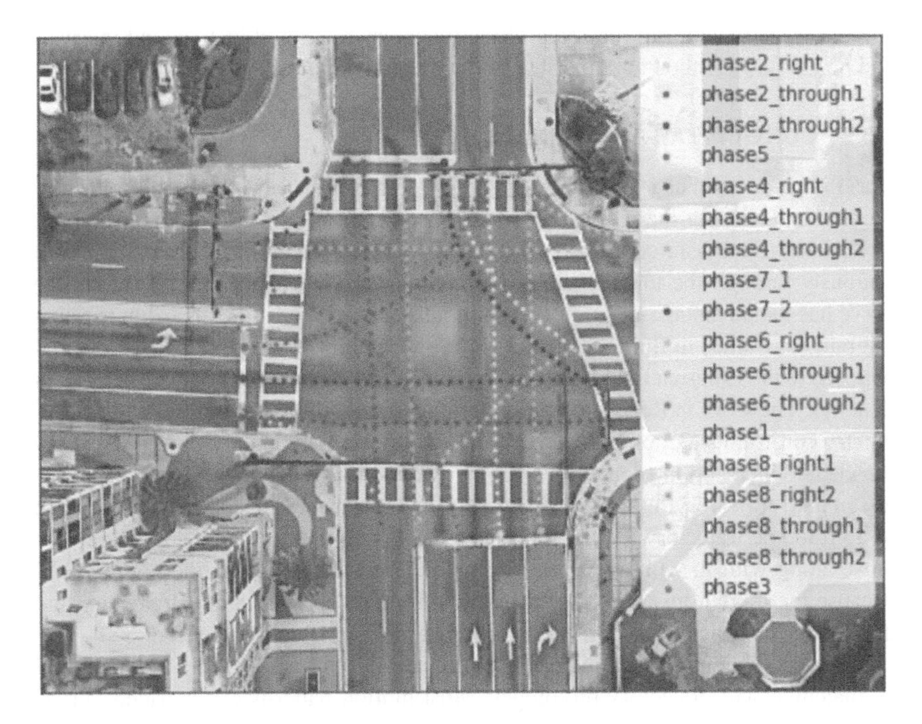

Figure 6.4 The resulting prototypes. Note that the slight deviation of some prototypes is caused by imperfect Google Map alignment

6.5 TRAJECTORY PREDICTION MODEL

In this section, we present our trajectory prediction model. A summary of the approach follows. The trajectories represented initially in cartesian coordinates are first transformed to CCS. These CCS trajectories are then fed to the LSTM-ED model for training and inference. In the following three sections, we define the trajectory prediction problem in Section 6.5.1, introduce the CCS and transformations from and to cartesian coordinates in Section 6.5.2, and describe the architecture, training, and inference using LSTM-ED in Section 6.5.3.

6.5.1 PROBLEM DEFINITION

Given an observed partial trajectory of the ith vehicle in CCS:

$$X_i = [(x_i^{(1)}, y_i^{(1)}), (x_i^{(2)}, y_i^{(2)}), \ldots, (x_i^{(t_{obs})}, y_i^{(t_{obs})})], \tag{6.3}$$

the model predicts the future trajectory:

$$\begin{aligned} Y_i = [&(x_i^{(t_{obs}+1)}, y_i^{(t_{obs}+1)}), (x_i^{(t_{obs}+2)}, y_i^{(t_{obs}+2)}), \\ &\ldots, (x_i^{(t_{obs}+t_{pred})}, y_i^{(t_{obs}+t_{pred})})], \end{aligned} \tag{6.4}$$

where t_{obs} is the current time step of the trajectory measured in 1-second intervals and $t_{obs} + t_{pred}$ is the final time step of the predicted trajectory.

6.5.2 CURVILINEAR COORDINATE SYSTEM

Curvilinear coordinate systems (CCSs) are a natural fit for our problem of trajectory prediction. Consider the problem of predicting left-turn trajectories at intersections. In standard Euclidean coordinates, the velocity vector changes its orientation continuously through the turn, whereas a reparameterization of the velocity along the curve has the advantage of better inertia representation. CCSs are also akin to using Lagrangian frames in fluid mechanics (or the viewpoint from the boat in a river) as opposed to the Eulerian frame (or the viewpoint from the riverbank watching the boat float by). The work in Ref. [146] uses curvilinear coordinates to impose roadway geometry constraints to motion tracking and behavior reasoning algorithms. We extend the idea to intersection geometry, which implicitly constrains the vehicle trajectories to standard trajectory templates. We propose to use the CCS, defined as the shape of a prototype trajectory. Partial trajectories matching the prototype are assumed to move along with it, with an offset. The prediction is largely simplified in the CCS because the model only needs to learn the difference between a new trajectory and the average historical trajectories conforming to the same motion pattern.

For a vehicle entering an intersection, the possible motion patterns can be found by matching with trajectory prototypes. Most motion patterns can be ruled out, as their distances to the query trajectory are too far to be considered potential matches. In this work, we only consider the closest two prototype trajectories. Each prototype trajectory defines a curvilinear coordinate system with s and n axes – essentially the tangent and normal at each point along the curve. The s-axis is defined along the shape of a prototype, with arc length coordinates used, while the n-axis is defined to be perpendicular to the s-axis at every point on the curve, giving us a measure of how far a point is from the curve, as shown in Figure 6.5. In the rest of this section, we describe the transformations between the image coordinate system (ICS) and the CCS, where image coordinates (IC) refer to Google Maps–aligned intersection image coordinates. We denote a point in IC as (x_p, y_p), and its corresponding point in CC as (s_p, n_p), as shown in Figure 6.5.

6.5.2.1 ICS to CCS Transformation

Our prototype trajectories are sampled from continuous splines, as in Section 6.4.2.2, and determining the closest point on a spline from a query point requires using optimization methods. These are now described. Given p with coordinates x_p, y_p in the ICS, we aim to find the CCS's corresponding s_p, n_p. We first find s_p by finding the closest point on spline S to the point p, which can be formulated as the following minimization problem:

$$\min_{s_p} ||p - S(s_p)||_2. \tag{6.5}$$

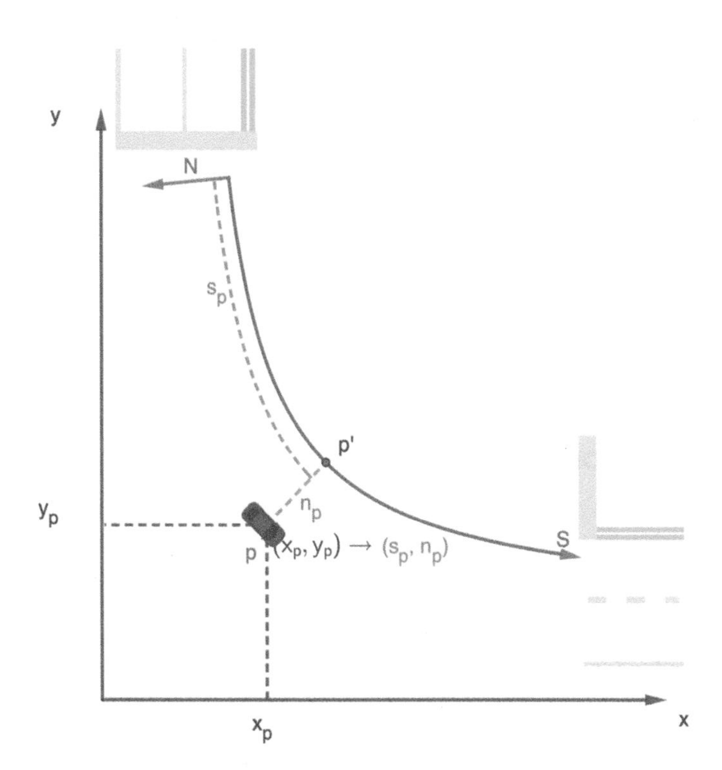

Figure 6.5 Illustration of curvilinear coordinate system

We use a standard limited-memory optimization algorithm (BFGS; available in Python scientific libraries) to solve for s_p. Once s_p is determined, we turn our attention to n_p. First, $|n_p|$ is defined as the distance between p and p'. The sign of n_p is determined by

$$\begin{aligned} &\operatorname{sgn}(\boldsymbol{pp'} \times \boldsymbol{q'p'}), \\ &q' = S(s_p - \epsilon), \end{aligned} \tag{6.6}$$

where sgn is the sign of a number, $\times$ is the cross product, q' is a point near p' with a slightly smaller s-value and ϵ denotes a small number. Vector $\boldsymbol{q'p'}$ estimates the growing direction of S at point p'. So far, we have given the procedure of transforming from ICS to CCS. We now examine the opposite direction.

6.5.2.2 CCS to ICS transformation

Given s_p, n_p in the CCS, we aim to find the corresponding x_p, y_p in the ICS. This is a relatively easy process because the transformation has a convenient closed-form expression:

$$\begin{aligned}(x_p, y_p) &= S(s_p) + \boldsymbol{p'p}, \\ \boldsymbol{p'p} &= n_p \boldsymbol{e}_\perp,\end{aligned} \tag{6.7}$$

where $\boldsymbol{e}_\perp$ is the unit vector perpendicular to S at point p', which can be found from the derivative of S. In summary, the use of CCS simplifies the task for trajectory prediction because it decouples the two coordinates to some extent. The S-coordinate mainly captures the speed along the road while the N-coordinate mainly captures the speed off the road (e.g., lane changes or abnormal behavior). The coupling of prototype trajectories to a "Lagrangian" coordinate system is a vital contribution of this work.

6.5.3 LSTM ENCODER–DECODER MODEL

LSTM networks are designed and proven effective for sequence modeling and prediction tasks. Thus, they are well-suited for trajectories represented as a sequence of coordinates. We adopt the encoder–decoder architecture to cope with variable lengths of trajectories and prediction periods. The encoder encodes observed partial trajectories to a fixed-length internal representation, and the decoder decodes the state and predicts possible future motions for a given period. Besides the internal vector from the encoder, cluster belonging is also provided as input to the decoder. In this way, the decoder learns to predict differently considering its cluster. Figure 6.6 illustrates our LSTM encoder-decoder training and inference behavior.

6.5.3.1 Network Architecture

The encoder and decoder have two layers: fully connected (FC) and LSTM layers. The FC layer is an embedding function that embeds locations into a fixed-length vector. The embedding will then be fed to the LSTM layer. The encoder encodes the observed trajectory into the LSTM's last hidden state $\boldsymbol{h}_i^{(t_{obs})}$. The decoder takes $\boldsymbol{h}_i^{(t_{obs})}$ concatenated with the one-hot-encoded cluster class vector as input and is trained to generate its future trajectory $\boldsymbol{Y}_i$.

6.5.3.2 Training

We adopt the L_2 loss for training, which measures the distance between the predicted and the ground-truth trajectories. As vehicles pass an intersection at different speeds, the number of trajectory points captured in the intersection varies over a wide range. For example, a *left-turn* trajectory usually has more trajectory points captured than a *straight-heading* trajectory because a *left turn* vehicle will slow down as it enters the intersection. In contrast, a *straight-heading* vehicle tends to be at the maximum speed limit. For this reason, we enable the model to encode variable-length observations and decode variable-length predictions. We achieve this by splitting a training sample into an observation and prediction sequence with a random ratio t_{obs}:t_{pred} so that the model is trained from mixed-length data and will learn to make variable-length predictions.

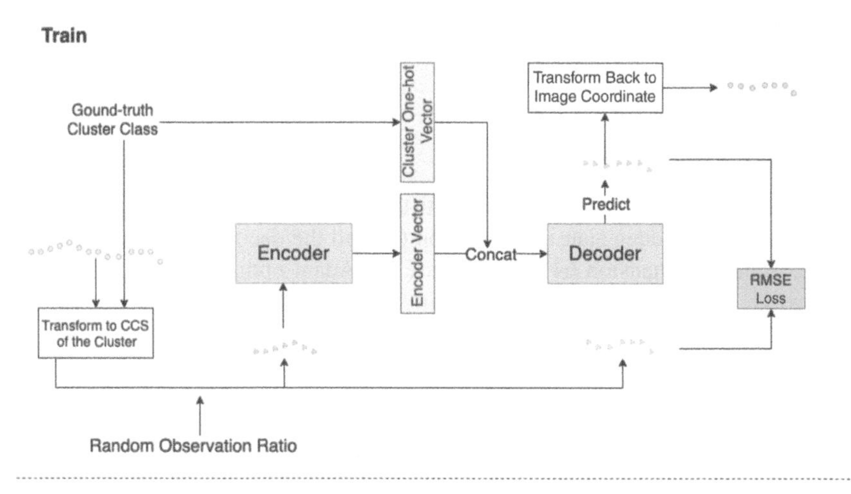

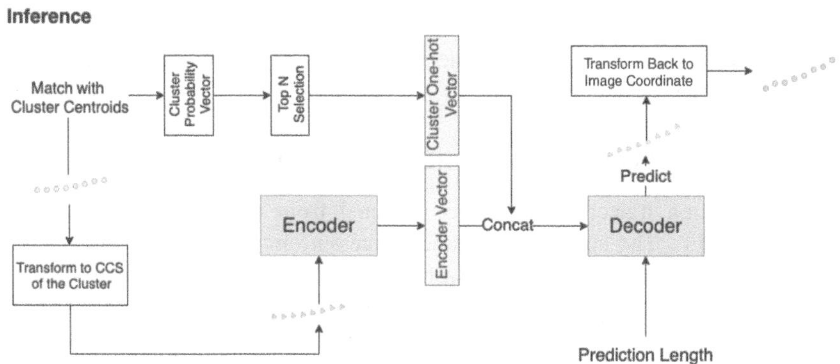

Figure 6.6 LSTM encoder–decoder training and inference behavior

6.5.3.3 Inference

At inference time, the cluster class of a trajectory i is unknown. The model first infers cluster class by matching with all prototype trajectories. The degree of belonging of trajectory tr_i to cluster class $C^{(m)}$ is calculated by inverse distance weighting:

$$u_{C^{(m)}}(tr_i) = \frac{1/d(tr_i, tr_{C^{(m)}})}{\sum_{j=1}^{n^c} 1/d(tr_i, tr_{C^{(j)}})}, \tag{6.8}$$

where d is the distance measure introduced in Section 3.5.1, and $tr_{C^{(j)}}$ represents the prototype trajectory of class $C^{(j)}$. If the trajectory has a similar degree of belonging to M classes, the model will output M predictions. For each possible cluster class, the model will make the corresponding prediction. To be more specific, for each possible $C^{(j)}$, the corresponding representation in CCS as well as the one-hot cluster vector, will be fed to the LSTM-ED model, which then makes its future prediction with "probability" $u_{C^{(j)}}(tr_i)$. We set M to 2 based on intersection geometry because each lane of an intersection usually allows two or fewer motion patterns (e.g., heading straight and right turn).

6.5.3.4 Implementation Detail

The embedding dimension is set to 64 for both the encoder and decoder. The encoder's LSTM layer has a hidden state of dimension 64, and the decoder's LSTM layer has a hidden state of dimension 64 + n_c. We use the Adam optimizer with a 1×10^{-4} learning rate. The t_{obs}:t_{pred} ratio is randomly chosen in the range $[0.3, 0.7]$.

6.6 EXPERIMENTS

We evaluated the proposed approaches on the collected trajectory dataset from surveillance fisheye cameras at three intersections. Section 6.6.1 describes our data collection methods and preprocessing steps to obtain Google Maps–aligned trajectories. Section 6.6.2 compared our trajectory prediction model with baseline methods. Finally, Section 6.6.3 evaluates the trajectory prediction pipeline for variable-length observation and prediction period.

6.6.1 DATA COLLECTION AND PREPROCESSING

Our intersection trajectory dataset is obtained from three busy intersections in Gainesville, FL. The surveillance cameras have a circular fisheye lens and a 10-fps frame rate. From our previous work [68], we have the mapping from fisheye video pixels to Google Maps locations. A detection and tracking pipeline is running periodically to automate the trajectory capture process. We clean the dataset using rule-based filtering. We impose speed limits within the intersection and non-self-intersecting constraints on trajectories. We also applied a trajectory smoothing algorithm to compensate for detection imprecision. In addition, as the traffic signal phase is unknown in our setting, we exclude the stay points of trajectories before entering the intersection. After the above process, we obtain roughly 15,000 trajectories for each intersection, split into training, validation, and testing sets with a roughly 7:1:2 ratio. The validation set is used in LSTM-ED training to avoid overfitting. We refer to the three intersections simply as Intersection 1, Intersection 2, and Intersection 3.

6.6.2 EVALUATION

In this section, we evaluate the performance of the proposed trajectory prediction model and compare it with baseline models. This experiment is performed on Intersection 1, and we assume the ground-truth cluster class of each trajectory is known. We compare the proposed model (LSTM-ED CCS) with two baseline models:

1. *LSTM-ED ICS:* An LSTM encoder–decoder model without coordinate transformation, the same as LSTM-ED CCS.
2. *GP:* A Gaussian process regression model used in Ref. [137].

Figure 6.7 Two predictions at inference time. The red, blue, and green points represent observed, ground-truth future, and predicted points of a trajectory, respectively

We adopt the following two metrics for prediction evaluation:

1. *Average displacement error (ADE):* The average Euclidean distance between ground truth and overall trajectory points prediction.
2. *Final displacement error (FDE):* The Euclidean distance between the end positions of ground truth trajectory and predicted trajectory.

We show the prediction errors of different observation and prediction periods. We set a threshold for minimum trajectory length for evaluation to ensure fair comparison for different period settings. Only trajectories with more than 40 timestamps are chosen because the maximum of observation and prediction timestamps is 40, as shown in Table 6.1. In this way, the same set of trajectories is used for each setting instead of some trajectories (less than 40 timestamps) only used in shorter-period settings. Thus, the experimental result truly reflects the model's performance for different observation and prediction periods. The prediction errors are reported in Table 6.1.

LSTM-ED CCS outperforms the two baseline models in almost every setting. GP tends to have a large FDE, and the prediction result also looks unsmooth. LSTM-ED ICS performs similarly on shorter prediction periods as LSTM-ED CCS but worse on more extended prediction periods.

6.6.3 EVALUATION OF TRAJECTORY PREDICTION PIPELINE

Our pipeline consists of cluster class determination and trajectory prediction. In other words, unlike the previous experiment, the ground-truth cluster class is unknown to the LSTM-ED model. We evaluated the pipeline at all three intersections. The quantitative result is reported in Table 6.2. The qualitative result is shown in Figure 6.8. As explained in Section 6.5.3.3, our model produces multiple outputs if the observed trajectory matches with numerous prototypes. Figure 6.7 shows one example of multiple outputs. From the observed trajectory, *going straight* and *right turn* are both likely.

Figure 6.8 Samples of the prediction results on Intersections 1, 2, and 3 (top to bottom, respectively)

Table 6.1

Comparison of Trajectory Prediction Approaches Given Ground-Truth Cluster Class

Observation length		10			20		30
Prediction length		10	20	30	10	20	10
GP	ADE	0.75	1.86	2.76	0.45	1.20	0.35
	FDE	2.76	4.96	6.08	2.34	4.58	2.69
LSTM-ED ICS	ADE	0.61	1.24	1.96	0.47	0.91	0.45
	FDE	1.13	2.51	4.17	0.86	1.97	0.84
LSTM-ED CCS	ADE	0.51	1.10	1.76	0.45	0.91	0.47
	FDE	0.97	2.32	3.61	0.81	1.87	0.87

* We utilize two metrics: average displacement error (ADE) and final displacement error (FDE) (in meters) to compare the three approaches. The lower the error, the more accurate the approach.

Table 6.2

Prediction Errors for Intersections 1, 2, and 3

Observation length		10			20		30
Prediction length		10	20	30	10	20	10
Intersection 1	ADE	0.54	1.21	2.09	0.52	1.16	0.61
	FDE	1.02	2.70	4.83	0.97	2.61	1.15
Intersection 2	ADE	0.54	1.23	2.05	0.46	0.91	0.49
	FDE	1.03	2.72	4.29	0.79	1.90	0.82
Intersection 3	ADE	0.66	1.40	2.19	0.58	1.18	0.54
	FDE	1.24	2.89	4.38	1.03	2.44	0.99

6.7 CONCLUSION

A real-time trajectory prediction approach coupled with aligned Google Maps information is proposed in this chapter. Our approach uses a historical trajectory database and finds typical motion patterns to guide future prediction. Given this prior information, our approach can make reasonable predictions based on variable starting position, observation period, and prediction period. Experimental results on three intersections show the effectiveness and extensibility of our approach. In the immediate future, we plan to integrate our trajectory prediction module into an early warning system. We believe our work will help increase intersection safety.

7 Vehicle Tracking across Multiple Intersections

7.1 INTRODUCTION

Mitigating traffic congestion and improving safety are the essential cornerstones of transportation for smart cities. With growing urbanization worldwide, traffic congestion along high-volume signalized traffic corridors (arterials) is a significant concern. Congestion negatively affects productivity, leading to loss of work hours, thus impacting the economy. Congestion also impacts the well-being of society and the environment [4, 147].

One of the critical congestion measures is arterial travel time [148]. This value is a vehicle's expected travel time to complete its journey along a signalized traffic corridor. It is affected by many factors, such as traffic conditions and departure time. This quantity is easy to interpret by traffic engineers, city authorities, and the general public. Traffic engineers can use travel time to identify problematic locations and timing problems that can improve the overall performance of a traffic system [149, 150, 151, 152, 153, 154]. Current performance evaluations only include a limited comparison of before-and-after travel time data to demonstrate the effectiveness of signal retiming [155] efforts.

Most existing research on travel time estimation focuses on using nonvision sensors such as GPS [156, 157] and loop detectors [158]. Due to the simplicity of implementation and low computational cost, travel time estimation methods based on historical data have been widely used in practice [159, 160]. Other approaches rely on machine learning and data mining of toll collection information [161], probe car [162], highways [163], and trip datasets [164].

However, traffic patterns vary dynamically during the day and globally within the network. There is a need for continuous monitoring and evaluation of signal timing parameters based on performance and fluctuation demands. Travel times have to be calculated at regular intervals. In addition, it is essential to understand the distribution of travel times rather than average travel times because the tail of the distribution gives traffic engineers rich information. Actual travel time often has a multimodal distribution, and the average values are not always sufficient. The advent of video cameras at traffic intersections has opened the possibility of using them for instantaneous travel time computation and is the focus of this chapter.

The novelty of our travel time estimation is twofold. First, we use car signatures to estimate vehicle departure and arrival time from video across multiple cameras. In addition, we leverage signal data in this system obtained from ATSPM logs from the controller (Figure 7.1). To our best knowledge, this represents the first end-to-end framework that estimates vehicle travel time on corridor intersections using multiple video and signal data.

DOI: 10.1201/9781003431176-7

Figure 7.1 An illustration of multi-camera vehicle tracking and travel time estimation problem. In this example, we have three intersections, A, B, and C, in a corridor. They all have one fisheye camera, and A and B have another pan camera. The multi-camera vehicle tracking and travel time estimation problem determine (1) whether the red car is the same vehicle among these intersections and (2) what its arrival and departure times are at each intersection

We achieve the above goals by accurately tracking a small subset of vehicles on a corridor or network. Along with reasonably synchronized clocks at each intersection, this will allow for accurate travel time computation for that subset of cars and can be used for travel time computation in general. The latter is a relatively straightforward computation if tracking can be accurately achieved. It is worth noting that it is more critical for the tracking to be accurate for a subset of cars rather than tracking a larger fraction of vehicles with less accuracy. That is because precise tracking can provide good input for reidentifying cars across multiple cameras and thus compute more accurate travel time.

We develop a unified system to track vehicles across multiple video cameras on corridor intersections. It uses a robust signature matching algorithm to handle the variance of vehicle pose under different camera perspectives. This algorithm can efficiently identify signatures of vehicles in real time and further help the computation of accurate detection of arrival and departure times at each intersection. Some vehicles may have to wait when the light turns red at any signal. This, in general, creates additional delays for a subset of vehicles. Our approach can robustly tackle these challenges.

Additionally, our approach leverages the fact that ordering vehicles from one intersection to another remains relatively the same. This is optional for our approach to be practical, but this property can reduce the computation time requirements for the matching algorithm. Although much of the chapter is described in terms of computing travel times on a corridor, the basic ideas are promising in the extension to work on a mesh of intersections.

In summary, we propose a real-time video processing system for multi-camera vehicle tracking and travel time estimation with the following key contributions:

1. We propose an exact video-based signature reidentification method crucial in multi-camera vehicle tracking.
2. We introduce a novel way to extend pairwise signature matching to phase-based signature matching to tackle the sequential group signature reidentification problem on video data.
3. We propose a novel travel time estimation method using arrival and departure information at each intersection.
4. We evaluate the proposed framework on a novel dataset from intersections in Florida with pan and fisheye cameras. The experimental results demonstrate promising performance for camera signature reidentification and travel time estimation.

7.2 METHODOLOGY

Multi-object multi-camera tracking intends to detect and track multiple objects within one camera (tracklets) and then perform multi-camera tracklet matching to derive trajectories of vehicles. The arrival and departure timestamps of the vehicles are then utilized to estimate travel time distribution. The inputs of our method are traffic video data collected from multiple intersections. For the sake of simplicity, we assume a three-camera system with one camera for each of the three successive intersections (corridor intersections). It is easy to generalize these ideas to a more significant number of intersections (e.g., mesh intersections). Each intersection can have one or more cameras, potentially of different types. In particular, the corridor that we experimented with has three intersections, labeled A, B, and C. A and B have one fisheye camera and one pan camera, while C has only one fisheye camera. Thus, this corridor corresponds to two tracking channels: (1) in the fisheye channel, the inputs are video sequences from three fisheye cameras, and (2) in the pan channel, the inputs are video sequences from two pan cameras.

Our proposed method contains four major modules, as shown in Figure 7.2:

1. *Multi-object Single-camera Tracking Module:* We follow the tracking-by-detection paradigm to track vehicles (car, bus, and truck) in every camera and generate local tracklets.
2. *Video-based Signature Reidentification (ReID) Module:* We parse our local tracklets into signatures in different cameras with departure timestamps and use a deep learning–based discriminative model to match the local tracklets across multiple cameras.

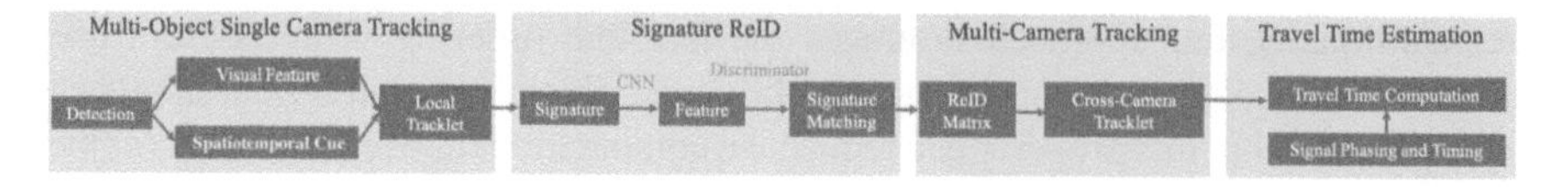

Figure 7.2 An illustration of the pipeline for the proposed method. Given intersection videos as input, a single-camera tracker first detects vehicles and generates local tracklets. Then a ReID discriminator computes matching among vehicles under the constraints of phase information. Finally, we apply a merging algorithm to update tracking results for multi-camera tracking and compute the travel time of each vehicle using the timestamp information

3. *Vehicle Tracking Module:* We leverage the signature matching results to associate the local tracklets (full timestamps) across multiple cameras by tracklet-to-signature assignment. By reconnecting the split local tracklets among different cameras, each vehicle obtains a complete trajectory across all cameras in multiple road segments.
4. *Travel Time Estimation Module:* We compute travel time based on arrival and departure timestamps from final across-camera tracking results. We assume that the input videos are collected nearly simultaneously and that local clock time can be converted to a common global time.

Additionally, before utilizing the second video-based signature ReID module, we can filter out tracklets using trajectory direction detection (go-straight, turn-left, or turn-right) and only focus on one direction (e.g., traffic flow go-straight from camera A to camera B and then camera C).

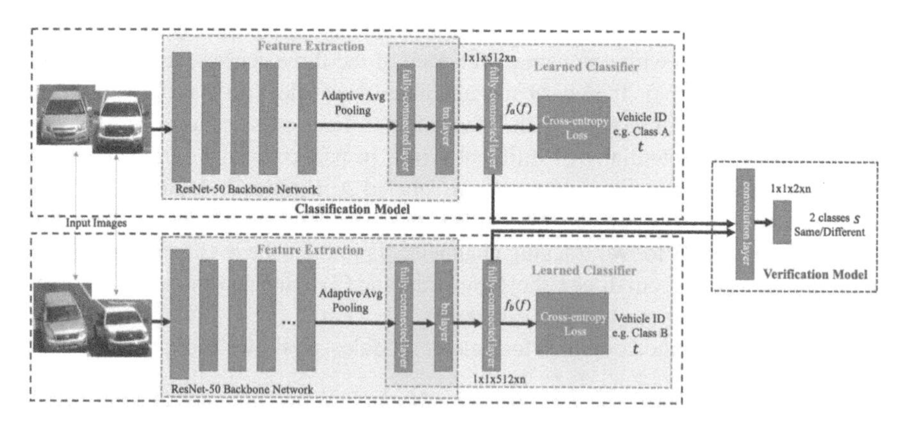

Figure 7.3 An illustration of the network architecture of our proposed two-loss signature ReID model using ResNet-50 as the backbone

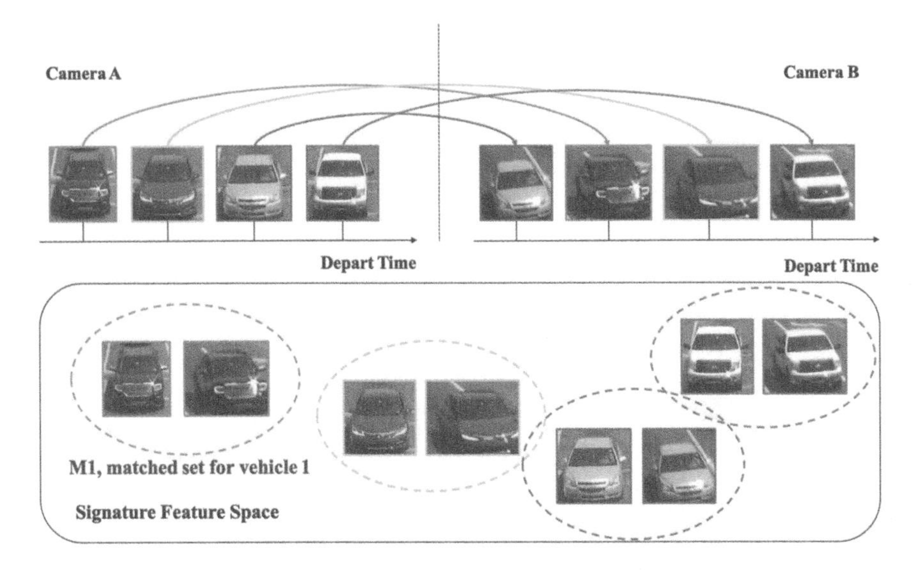

Figure 7.4 An illustration of matched-vehicle sets and their distributions in the feature space. The solid color arrows indicate the one-to-one mapping results

Each of these modules is described in detail in the following sections.

7.2.1 MULTI-OBJECT SINGLE-CAMERA TRACKING

We use a deep learning model to detect and track road objects, calculate the speed after deformation correction, and then calculate the map-based trajectory. The deep object detector trained on fisheye video samples is based on the architecture of YOLO [47]. According to the intersection attributes, we specify five object categories: pedestrians, motorcycles, cars, buses, and trucks.

The multi-object tracker is built on DeepSORT [52] and uses the conventional single-hypothesis tracking method with recursive Kalman filtering, reference, and frame-by-frame data association. However, when the intersection becomes crowded or large buses or trucks appear, there is an occlusion problem. Therefore, some road objects can obtain new recognition after the occlusion disappears, forcing us to integrate object signatures or ReID features.

We introduce a deep cosine metric learning component to learn the cosine distance between road objects and integrate it as the second metric for the association problem in multi-object tracking [73]. The cosine distance includes the appearance information of the road object to provide valuable hints to restore the identity when the discriminative power of the motion feature is small. We trained a deep cosine metric learning model on the VeRi dataset [165]. To ensure we generate good inputs for ReID, we introduce a track direction detection component to focus on vehicles that go straight from camera A to camera B and then to camera C. Figure 7.4 shows

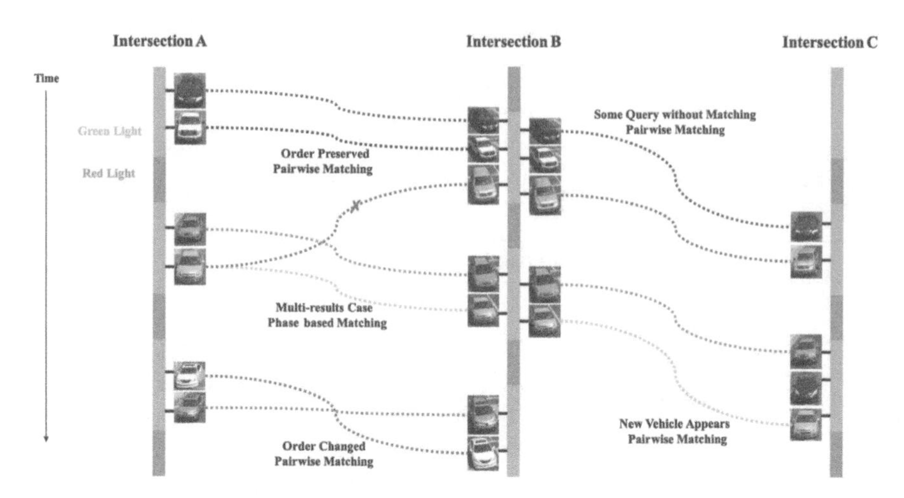

Figure 7.5 An illustration of across-camera matching. Our primary matching strategy is pairwise matching using our discriminator. For cases with multiple matching results, phase information is used as an additional constraint to refine matching. We defined five types of matching cases: (1) order preserved (pairwise matching), where the order of signature sequence does not change among intersections; (2) multi-results case (phase-based matching); (3) order changed (pairwise matching); (4) some query without matching (pairwise matching); and (5) new vehicle appears (pairwise matching)

the matched vehicle sets and their distributions in the feature space. The direction detection is based on the computation of vehicle velocity direction from trajectory data. The signal data is used before ReID to group local tracklets. Figure 7.5 is an illustration of across-camera matching. We divide tracks into different phases based on green light timestamps.

7.2.2 PAIRWISE SIGNATURE REID

Vehicle signature ReID is used to determine whether a specific vehicle is across images or video frames from nonoverlapping cameras. In real-world scenarios, vehicles have both rigid and flexible characteristics due to the differences between different camera settings (camera type, perspectives, height, FOV, etc.). Their appearance has constraints within the range of color, size, and type, and the appearance is easily affected by occlusion and viewing angle, which makes vehicle ReID a challenging task. The popular deep learning ReID methods leverage high-level semantic information and design novel objective functions to encourage models to learn a discriminative feature representation. The two main types of loss functions are classification loss and verification loss. Classification loss directly uses the label information of the signature to perform multiclass classification. The verification loss determines whether the two input signatures belong to the same object. Hybrid approaches have improved performance by combining both in model learning [166]. Therefore, we choose to use both classification loss and verification loss.

7.2.2.1 Overall Network

Our signature ReID model uses deep learning models to obtain discriminative feature representation. Figure 7.3 shows the overall model architecture of our signature ReID network. Our network consists of two submodels – the classification and verification models. Given the vehicle's image as input, the classification model distinguishes images by dividing them into N classes based on which vehicle they belong to, which is learned by a softmax loss. For example, we assume a total of N objects presented in the training dataset during training. We therefore use an N-class softmax classifer to distinguish these objects. However, when moving to inference, the objects are completely changed; thus, the N-class classifier is no longer helpful. As our primary goal is to obtain the signature of each object rather than an actual classification, we instead use the features of the last fully connected layer (a feature vector of 512) before the softmax classifier as its signature feature. This makes the model applicable to both training and inference for any object. Given a pair of vehicle images, the verification model determines whether the two images come from the same object, by mixing their corresponding 512-dim embedding features via element-wise multiplication and stacking several convolutional layers to form one binary classifier. We use the 512-dim fully connected feature as a vehicle-signature descriptor. If two vehicle images come from the exact vehicle (having the same class ID), we classify them as 1; otherwise, they are 0. The binary classification will encourage the 512-dim signature features to be similar if belonging to the same object and more distinct if coming from different objects.

Since signature ReID is the key to multi-camera tracking, our ReID discriminator must achieve high precision to find the best signature matching. Given a pair of images, we apply a Siamese network to calculate the classification losses for both images and compute the verification loss between the two, which simultaneously predicts the IDs of the two images and the similarity score. It is supervised by the classification label c and the verification label v.

7.2.2.2 Classification Loss

In the Siamese architecture, the two ResNet-50 models share weights in the network framework and predict the category labels of the pair image. The original model's fully connected layers and the last pooling layer are removed. We use the adaptive average pooling to obtain a feature with a fixed dimension, followed by several new fully connected and batch norm layers. Before feeding to the final classifier, we can obtain a 512-dim fully connected feature (denoted as f), which serves as the vehicle signature descriptor. Because we train the model on the vehicle ReID dataset VeRi [165] with 576 vehicles, each with multiple images, the classifier consists of one fully connected layer with 576 dimensions (denoted as ϕ_θ with parameters θ)

and a softmax layer to obtain the class distribution. The cross-entropy loss is used for training the classifier, which is denoted as follows:

$$\hat{p} = softmax(\phi_\theta(f)), \tag{7.1}$$

$$\text{Identif}(f, t, \phi_\theta) = \sum_{i=1}^{K} -p_i \log(\hat{p}_i), \tag{7.2}$$

where t is the target class, $\hat{p}$ is the predicted probability, and p_i is the target probability. K is the total number of classes and is set to 576.

7.2.2.3 Verification Loss

The verification loss directly takes two descriptor vectors f_1 and f_2 computed from two images as inputs. We compute $f_s = f_1 * f_2$. After this, we stack several fully connected layers and add the softmax classifier to project f_s to a 2-dim feature vector, representing the predicted probability of whether the two images come from the same vehicle or not. It, therefore, regards the ReID problem as a binary classification problem where the cross-entropy loss function is similar to the classification model. The cross-entropy loss in the verification loss is as follows:

$$\hat{q} = softmax(\gamma_\theta(f_s)), \tag{7.3}$$

$$\text{Verif}(f_1, f_2, s, \gamma_\theta) = \sum_{i=1}^{2} -q_i \log(\hat{q}_i), \tag{7.4}$$

where f_1 and f_2 are two tensors of size $1 \times 1 \times 512$, s is the target category indicating the same or different target, and γ_θ denotes the whole operation that maps f_s to the 2-dim feature vector before the softmax layer.

7.2.2.4 Losses

The verification loss is directly trained on the similarity between two features through an intuitive pairwise comparison method. The disadvantage of the verification loss is that only the verification result is considered in training, while the annotation information is not fully utilized. Also, the association information between the image pairs and other images in the dataset is not utilized.

The classification loss regards the vehicle ReID task as a classification task; each vehicle signature with the same identity is regarded as a category. The classification loss learns directly from the input image and its identity ID. The inputs of the classification model are independent. Still, a potential association relationship is implied because each signature has an implicit relationship with a signature with the same identity and a signature with different identities through category tags. The biggest flaw in the classification model is that the training target is entirely different from the test process. The embedding features are extracted and used to compute the similarity during the test. However, the similarity measurement information between the image pairs is not considered during training.

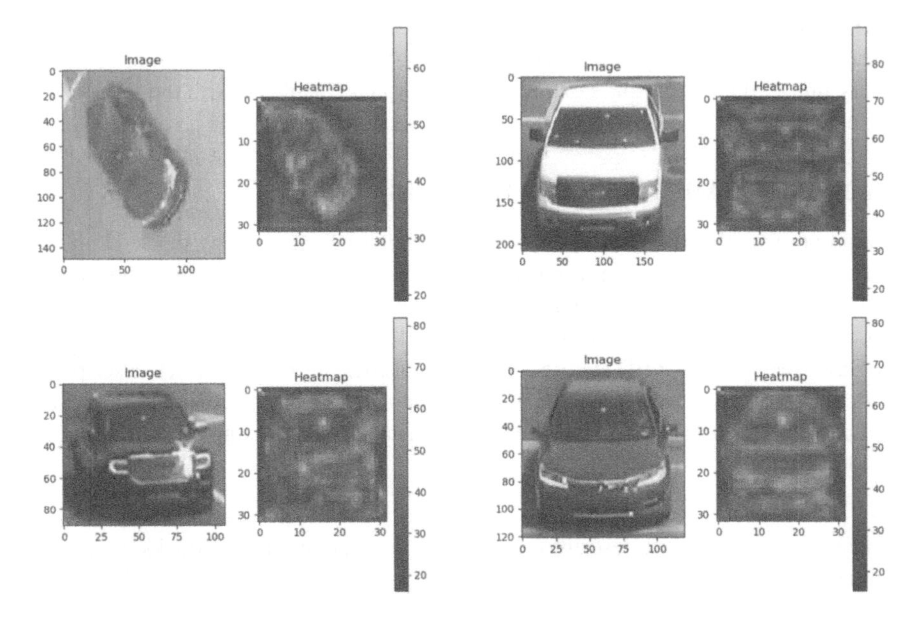

Figure 7.6 Samples of activation map of our model trained by the two losses

We have visualized the sum of several activation maps. The classification and verification networks show different activation patterns for vehicles. The ReID network often finds a discriminatory part if only one loss is used. Our proposed model takes advantage of the two networks, and the new activation map is mainly a combination of two separate maps. The proposed model enables more neurons to be activated. In Figure 7.6, we show some examples of the two-dimensional visualization of the embedding.

7.2.2.5 Training and Optimization

During training, we resize our input images to 256 × 128, shuffle the dataset, and use a random order of images. Then, we sample another image from the same (or different) classification ID to form a positive (or negative) pair. Initially, the ratio of the negative pair to the positive pair is 1:1. We gradually multiply the ratio to reduce the forecast bias until it reaches 1:4. This is beneficial because the number of positive pairs is minimal, and the network runs the risk of overfitting. We set the total training epochs to 75 and a batch size of 32. The training begins with an initial learning rate of 0.001 and decays the learning rate to 0.0001 in the last five epochs. Stochastic gradient descent (SGD) is used to update network parameters. The weight of the verification loss is set to 1, and the weight of the two classification losses is set to 0.5. The dropout layers are applied as well.

Given a 256 × 128 image in the testing phase, we feed it to our trained network and obtain the vehicle descriptor f. We obtain another descriptor, f_{flip}, by feeding its horizontal-flipped image. f and f_{flip} are averaged to form the final descriptor. After obtaining the descriptor of the candidate set (gallery set), we save it offline. For the query image, the descriptor is extracted instantly and calculated with the features of the candidate set (query set) to obtain the final matching result.

7.2.3 MULTI-CAMERA VEHICLE TRACKING

We propose an effective, fast multi-camera vehicle tracking strategy to speed up the matching process and obtain more stable synchronization results using intersection and other information. The signal data are used before ReID to group local tracklets, e.g., dividing tracks into phases based on green light timestamps.

Based on signature ReID features and temporal cues, we build our track descriptors with a bag of information: signature_match, camera_ID, timestamp, original_track_ID, and class (car, bus, etc.). We first compute the distance matrix using track descriptors as follows:

$$distance_N = [d_{i,j}]_{i,j=0}^{i,j=N} = 1 - cos(t_i, t_j), \tag{7.5}$$

where t_i and t_j are track descriptors, and N is the total number of tracks of single-camera tracking from all intersections. We set a very high value for vehicles in the same camera images for the distance matrix because they are not supposed to be merged. We use one predefined threshold for track merging.

Given the distance matrix $distance_N$, we update the previous multiple object tracking result T following these rules:

1. Sort the small traces according to their camera IDs and compare only small traces with adjacent camera IDs.
2. If the minimum distance between the query track descriptor and all other track descriptors under different cameras exceeds the merging threshold, the tracklet T_i is removed.
3. If the distance between q_i and T_i and q_j and T_j are all smaller than a predefined merging threshold, update the tracking ID of T_i and T_j to be the same.

Here T_i represents a gallery of small tracks with tracking function ID i and q_i represents small tracking queries with tracking ID i.

At intersections, vehicles can go in multiple directions. It is more efficient to detect the trajectory direction and then apply ReID and multi-camera tracking. Our dataset has three intersections, A, B, and C, in a corridor. They all have one fisheye camera, and A and B have another pan camera. We filter out tracklets using trajectory direction detection (go-straight, turn-left, or turn-right) and only focus on one direction (e.g., traffic flow go-straight from camera A to camera B then camera C).

Without loss of generalization, for the rest of this chapter, we focus on one-direction tracking for vehicles as they appear at intersection A, then intersection B, and then intersection C. We apply a simple direction detection method based on trajectory information and intersection topology.

7.2.4 TRAVEL TIME ESTIMATION

We assume the following:

1. The clocks used for each of the input videos are already synchronized. If they are not, it is easy to add the differences in times to our calculations.
2. We filter out tracklets using trajectory direction detection (go-straight, turn-left, or turn-right) and focus on only one direction (e.g., traffic flow go-straight from camera A to camera B and then camera C).

Since the videos are synchronized, we can compute travel time based on arrival and departure timestamps from final across-camera tracking results:

$$T_{camera_A,camera_B} V_i = timestamp_B V_i - timestamp_A V_i. \tag{7.6}$$

In the overall pipeline, the single-camera tracking generates tracklet results with information such as camera ID, video_start_time, frame ID, track ID, class, width, height, x, and y. The signal data provide phase information (red or green light), and we use it to group a subsequence of tracklets to one phase. So for signature ReID, the input is a signature representation that includes the image (cropped from detection bounding boxes), track ID, camera ID, frame (departure time), class, and phase ID.

The pairwise matching windows are about two phases that usually contain 4-min video data with about 20–30 vehicles for one direction per camera. The ReID model outputs the matching results under constraints: (1) the similarity score is higher than the threshold, (2) two signatures are from different camera IDs, and (3) if multiple matching is computed, compute the phase matching matrix (matching phase ID for all signatures in the same phase) and add a penalty for matching results where the matrix distance is larger. We obtain the matching results and use them as registration information for cross-camera tracking for multiple local tracklets. Finally, we have across-camera tracking results with information to compute travel time – start intersection, end intersection, signature ID, departure_time, and arrival_time. We also implemented a visualization part of travel time distribution to aid travel time analysis.

7.3 EXPERIMENTS

In this section, we first introduce our dataset and experiment settings. Then, we present qualitative and quantitative experimental results.

7.3.1 EXPERIMENTAL SETUP AND PARAMETER SETTING

The experiments were performed on a 256-GB RAM machine with 16 CPUs and 1 NVIDIA graphics card (Titian V). Our signature ReID was implemented on Pytorch. We train a ResNet-50-based signature discriminator on the VeRi dataset as the pre-trained model and evaluate our dataset collected from pan and fisheye cameras of corridor intersections in Florida. The VeRi dataset contains over 50,000 images of 575 vehicles captured by 20 cameras. The threshold we set for the signature discriminator is 0.7.

7.3.2 DATASET

The data we used in this method is pure traffic video data collected from three intersections in a corridor in Gainesville, FL. We refer to the three intersections as A, B, and C. A and B have one fisheye camera and one pan camera, while C has only one fisheye camera. The input of multi-object multi-camera tracking is the M video sequence from M cameras. Since we have two types of cameras, we have two tracking channels: (1) in the fisheye channel, the input is three video sequences from three fisheye cameras, and (2) in the pan channel, the input is two video sequences from two pan cameras.

We evaluate our method using a self-curated video dataset from Intersections A, B, and C. The resolution of pan camera video files is 1,280 × 720, and that of the fisheye camera video files is 1,280 × 960. The duration of the video files is about 16–20 minutes, containing approximately eight or nine phases. Our test set contains over 200 image sets with over 2,500 vehicle detections. The vehicle types of the test set include cars and buses. Based on single-camera tracking results, the signature of each vehicle consists of the detection image (cropped from bounding boxes), camera ID, frame ID, and track ID. Each vehicle has about 11 images (near departure time) per camera, and we can apply a multi-query matching setting by averaging these 11 images for signature ReID.

7.3.3 QUALITATIVE RESULTS

We present qualitative results of signature matching. We visualize the sum of several activation maps of features from our signature-matching network. As shown in Figure 7.6, the signature ReID network shows different vehicle activation patterns. In Figure 7.7, we present sample results of cross-camera tracking with estimated travel time. In this example, all vehicles that are going straight (left two lanes) in Intersection A (designated as Intersection 1 in the illustration) have correct signature matching at Intersection B (defined as Intersection 2 in the illustration). During this phase, for the group of vehicles (11 vehicles), the average travel time is about 57.3 seconds. We present examples of pairwise signature matching results in Figure 7.8, demonstrating good matching for both fisheye and pan camera data. In Figure 7.9, we show examples of the top ten matching results for queries from our and the VeRi datasets. Our gallery dataset includes pan and fisheye cameras to compute the top ten matching results. The experimental results show that our discriminator network has reasonably high accuracy in retrieving correct matching.

7.3.4 QUANTITATIVE RESULTS

To evaluate our signature ReID network, we first evaluate our test dataset with accuracy in terms of Rank 1, Rank 5, Rank 10, and mean average precision (mAP). The

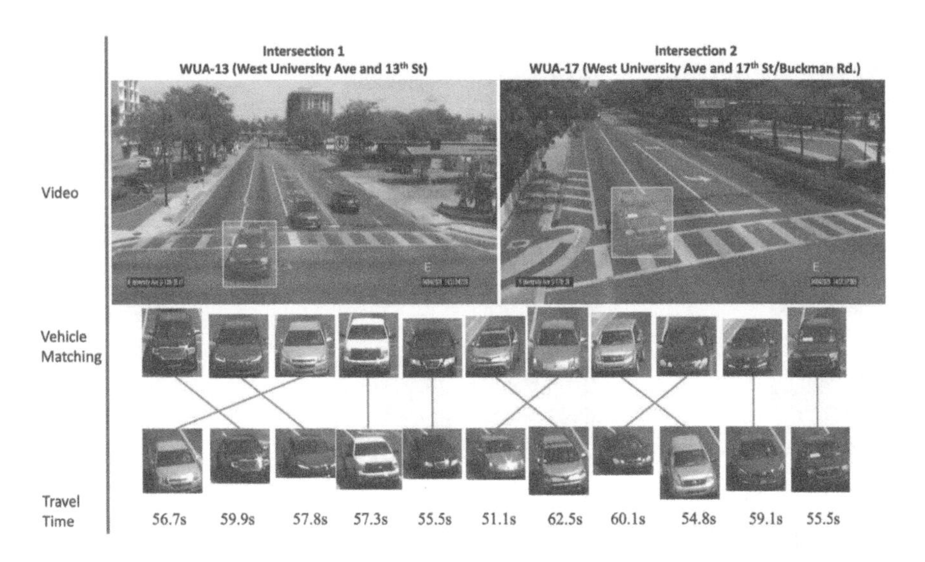

Figure 7.7 Multi-camera tracking and travel time estimation results. Results of one-phase data after first-order matching, sequence order checking, and travel time estimation. Multi-matching did not happen, but sequence order changes. The travel time between these two intersections ranges from 54 seconds to 62 seconds

definition of mAP is as follows:

$$\mathrm{mAP} = \frac{1}{N}\sum_{i=1}^{N}\sum_{j=1}^{K_{q_i}}\left(\hat{r_j}/r_j\right), \tag{7.7}$$

where query dataset is denoted as $Q = \{q_1, q_2, \ldots, q_i, \ldots, q_N\}$ with N images and gallery dataset is denoted as $G = \{g_1, g_2, \ldots, g_i, \ldots, g_M\}$ with M images. For each query q_i, we sort the gallery data in ascending order of ReID distance and denote the sorted gallery as G_{q_i} and the matching subset in G_{q_i} as $\hat{G}_{q_i} = \{\hat{g_2}, \hat{g_1}, \ldots, \hat{g_j}, \ldots\}$. Assuming $\hat{g_j}$ in G_{q_i} has index r_j and in $\hat{G}_{q_i}$ has index $\hat{r_j}$, we repeat this query processing for all query data. Rank 1 is defined as $rank1 = \hat{N}/N$ where q_i meets Rank 1 if the first match is correct in G_s of q_i. The number of queries that meet Rank 1 is Q is N.

We train our signature ReID model on 576 training sets from the VeRi dataset and test it on our dataset with 200 testing sets (including both pan and fisheye data). The test and training datasets were disjoint. Table 7.1 shows that we have outstanding accuracy in both single-query and multi-query settings in terms of Rank 1, Rank 5, and Rank 10. The mAP of 0.853039 for a single query is also good. A quantitative prediction of single-camera tracking, pairwise signature ReID, and phase-based signature ReID is achieved by comparing predicted results with the ground truth at the frame or object levels. We apply a predefined threshold (e.g., 0.7) to compute matching candidates and pick the highest score for the final results. A true positive is a match where

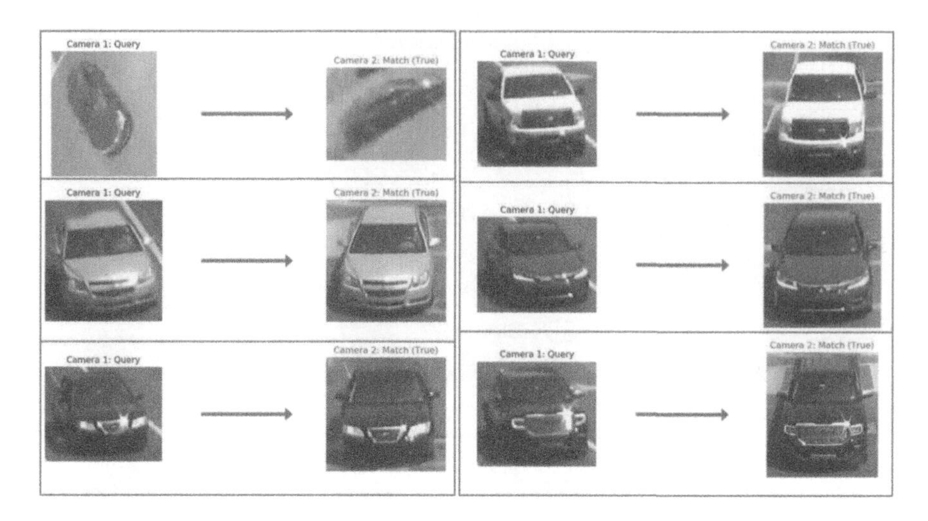

Figure 7.8 Samples of pairwise signature matching results

the object signature IDs are the same, but the camera IDs are not the same. A false positive occurs when the object signature IDs are not the same. A false negative occurs if there is no match with ground truth and matching results. The precision, recall, and F1 score are defined as follows:

$$TNR = \frac{TN}{TN + FP} = 1 - FPR \tag{7.8}$$

$$Precision = \frac{TP}{TP + FP}, \quad recall = \frac{TP}{TP + FN} \tag{7.9}$$

$$F1 = 2 \times \frac{Precision * recall}{Precision + recall} \tag{7.10}$$

The evaluation metrics we used for quantitative evaluation include precision, recall, F1 score, and speed for the four major modules of the proposed method. Table 7.2 shows that our pipeline has achieved promising performance on three modules – single-camera tracking, pairwise signature ReID, and phase-based signature ReID.

Table 7.1

Quantitative Evaluation of Our Discriminator for Signature ReID Matching

Component			Single Query (Reranking)				Multi-query	
Task	**Dataset**	**Loss**	**mAP**	**Rank 1**	**Rank 5**	**Rank 10**	**Rank 5**	**Rank 10**
ReID	VeRi	Classification	0.67	0.91	0.93	0.96	0.88	0.95
ReID	VeRi	Classification + Verification	0.72	0.94	0.96	0.98	0.94	0.97
ReID	Ours	Classification	0.79	0.89	0.89	0.92	0.85	0.85
ReID	Ours	Classification + Verification	0.85	1.00	1.00	1.00	0.96	1.00

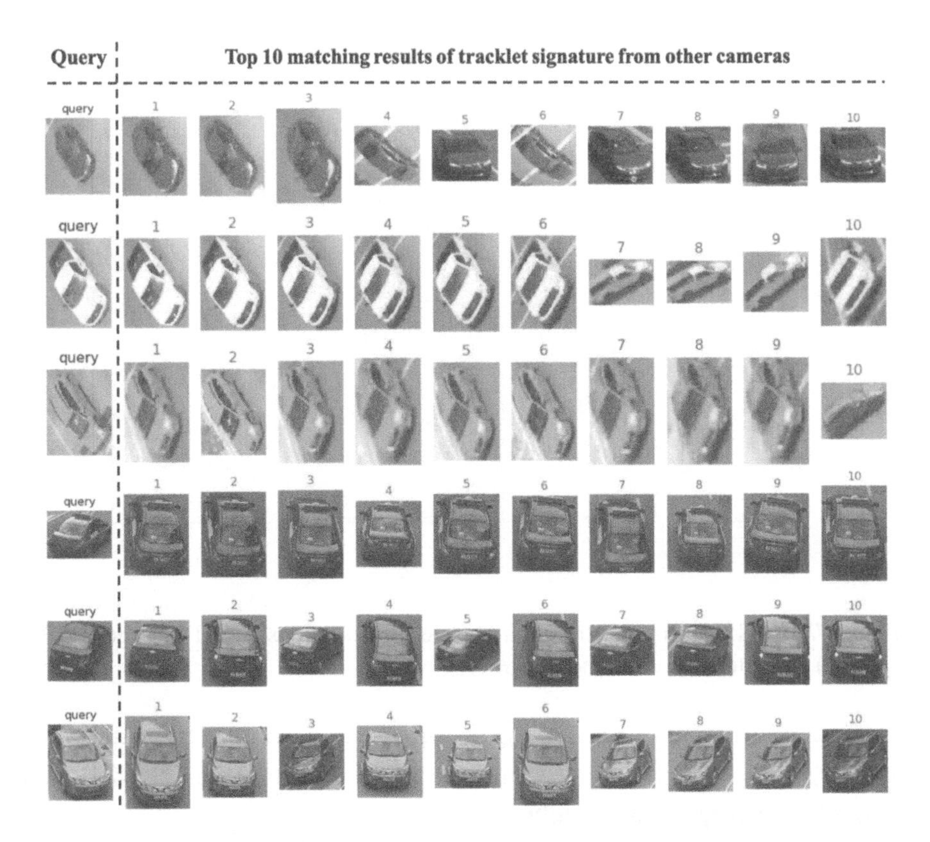

Figure 7.9 Samples of Top 10 matching results of signature ReID on our dataset (top 3 rows: pan + fisheye cameras) and VeRi dataset (bottom three rows)

Figure 7.10 and Table 7.3 show the test set's travel time distribution and statistics information. The distribution of travel times between Intersections A and B shows two spikes for this road segment. It aligns with our investigation – the first vehicle queued in the lane influences the actual travel time of that group of vehicles.

The signature ReID performance in both single-query and multi-query settings is shown in Table 7.1. The ReID model training takes about 2 hours on our device – one

Table 7.2
Quantitative Evaluation of Proposed Method

Task	Precision	Recall	F1 Score
Single-camera tracking	0.96002	0.93980	0.94980
Pair wise signature ReID	1.000000	0.836111	0.910741
Phase-based signature ReID	1.000000	0.816667	0.899082

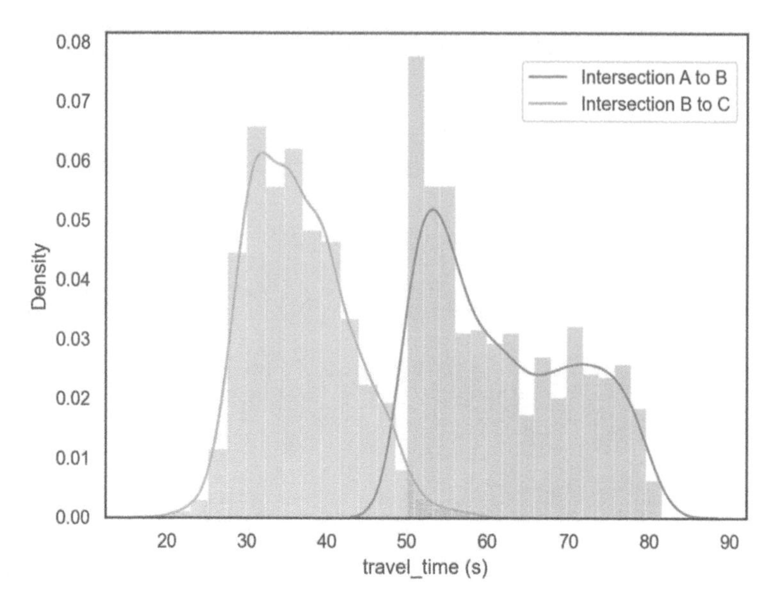

Figure 7.10 Travel time distribution between Intersections A and Intersection B (yellow) and Intersections B and C. (gray)

Table 7.3

Travel Time Computation Results

Intersections	Mean of Travel Times	Standard Deviation
A to B	61.80 s	9.03 s
B to C	36.55 s	6.18 s

Table 7.4

Processing Time Analysis of the Proposed Method for 5-minute Video Clip (300 seconds)

Component	CPU/GPU	Processing Time (seconds)
Single-camera tracking	NVIDIA TITAN V	166–170
Signature ReID	NVIDIA TITAN V	50–60
Travel time computation	CPU (16 cores)	5–7
Overall pipeline	GPU + CPU	221–237

NVIDIA Titan V. In Table 7.4, we present the processing time of each component for a 5-minute video clip (1,280 × 960, 10 fps). The overall pipeline costs 221–237 seconds and achieves real-time performance.

7.4 DISCUSSION

This chapter introduces a novel method and a real-time system to estimate vehicle travel time by leveraging video processing on multiple intersections.

Our key contributions can be summarized as follows: (1) We propose a novel video-based signature ReID method with high precision that plays a crucial role in multi-camera vehicle tracking. (2) We introduce a novel way to extend pairwise signature matching to phase-based signature matching to tackle the sequential group signature ReID problem on traffic video datasets. (3) We propose a real-time video processing system for multi-camera vehicle tracking and travel time estimation. We evaluated the proposed framework on a real-world dataset from intersections in Florida with pan and fisheye cameras. Although our results were presented for three intersections, the pairwise intersection nature of our approach allows for scaling this approach to a larger number of intersections. Overall, our experiments demonstrated the proposed approach's viability, effectiveness, and scalability. Our future work will extend this to computing origin–destination travel times to all pairs of inputs and outputs across all intersections in a corridor or a traffic network.

8 User Interface

8.1 INTRODUCTION

Most traffic intersections today have different sensors, such as video cameras, lidar units, and traditional induction loop detectors. Technological advances enable us to monitor these intersections remotely in real time and identify potential inefficiencies. However, the high volume of data being collected by the sensors at each intersection and the number of such intersections in a city makes it imperative to automate the process of monitoring and analyzing problem areas. Using state-of-the-art machine learning algorithms and data communication mechanisms at the back end, we have developed a visual analytics system to study the videos generated at an intersection.

The system can detect movement conflicts or near-misses in streaming mode within a small latency. The system also can perform historical analysis where the user specifies a time window of interest. The tool may filter and analyze the trajectories of vehicles and pedestrians passing through a traffic intersection during that time. The system is based on intersection videos obtained using fisheye cameras because fisheye cameras can capture a complete view of the intersection, unlike regular cameras that can record approaches from only one direction.

The main contributions of our work presented in this chapter are summarized as follows:

1. The tool allows users to filter trajectories based on an intersection's time and signaling phases. The filtered trajectories, in combination with the ongoing signal data, help the user to determine the trajectories that violate the red light.
2. The tool uses powerful clustering algorithms that segment the trajectories based on object type and the path traversed, resulting in interesting visualization-based insights of traffic behavior in terms of normal and safe versus rare and accident-prone. Besides, clustering also helps us determine counts of vehicles on different lanes as the trajectories on separate lanes form separate clusters.
3. The tool can be used to perform traffic analysis to get insights into key trends at an intersection. Visualizing the dominant traffic behavior in different settings may help the traffic engineer address potential bottlenecks by changing the timing plan or revamping the intersection.
4. The tool has streaming mode reporting capabilities, which is immensely beneficial for observing the traffic trends of an intersection with a small latency (a few minutes).
5. The tool maps the vehicles from a spherical coordinate system corresponding to the fisheye camera to a rectangular space. This can then be aligned with a mapping tool such as Google Maps.

DOI: 10.1201/9781003431176-8

A user interface (UI) was developed in Ref. [70]. Apart from the fact that the infrastructure is completely different for the two systems (the new system is Web-based, while the system in Ref. [70] is based on Qt), we have incorporated many new features in the new system. Some of these features are summarized here:

1. The new system and UI have two modes: streaming and historical. The streaming mode processes trajectories generated in small batches with a constant latency that the user may specify. The version presented in the conference paper was limited to historical mode.
2. We applied an unwarping algorithm to the images of the trajectories captured using a fisheye camera and represented them in rectilinear coordinates.
3. We present an extensive evaluation of the trajectory statistics via the new user interface. These statistics include turn movement counts, gap analysis, and speed of the trajectories.
4. We added support to display the centroids for the clusters of trajectories.
5. We added support to display a heatmap of the intersection point to areas more susceptible to incidents.
6. The user can click a trajectory, and then a video clip of the actual vehicle traversing the trajectory pops up if the video being analyzed is available in the file system.

The rest of the chapter is organized as follows. Section 8.2 presents the related work in trajectory analysis. Section 8.3 gives preliminary background on video processing. Section 8.4 shows the core of our visualization software. Section 8.4.2 is devoted to the historical analysis mode of our visualization software, and Section 8.5 presents a trend analysis case study. We finally conclude in Section 8.6.

8.2 RELATED WORK

The field of trajectory analysis for pedestrians and vehicles has become a hot area of research after the advent of intersection cameras. However, only some existing works have developed a complete analysis framework and a user interface for analyzing the videos. In this section, we mention some work that closely resembles our work in this book regarding the components that build the analysis system. Sha et al. [167] performed trajectory analysis using laser data obtained from objects using an intersection. Xu et al. [168] present an algorithm for clustering the trajectories at an intersection. In our work, we have developed visual analytics software on top of the analysis algorithms.

TrajAnalytics [169] is an open-source visual analytics software developed by Dohuki et al. TrajAnalytics is used for exploring urban trajectories and has built-in algorithms to perform modeling, transformation, and visualization of urban trajectory data. A fundamental difference between TrajAnalytics and our visual analytics system is that the software analyzes the whole trajectory of vehicles and pedestrians in the former. In contrast, we analyze only trajectory segments that appear at traffic intersections.

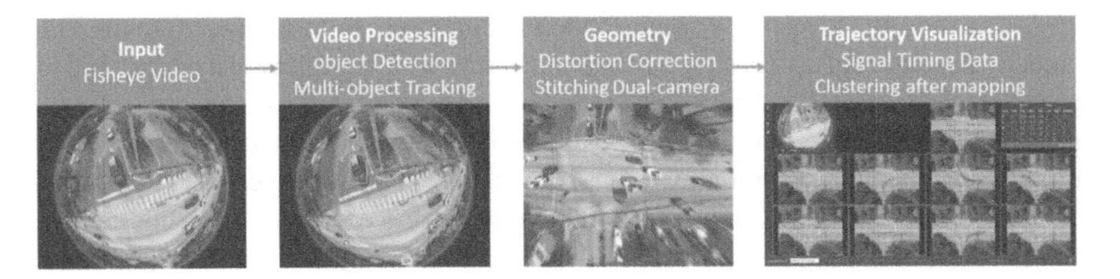

Figure 8.1 An overview of the pipeline consisting of video processing and multi-object tracking for trajectory generation, followed by post-processing and fusion with signal data and, finally, visualization

Another existing tool developed by Kim et al. [170] performs analysis at different stages, such as map view, trajectory view, analysis view, and timeline view. The tool provides the user options to filter areas of interest in the map view, generate the trajectories of interest, and study the timeline of these trajectories. This work also analyzes entire trajectories from start to end, and hence their data collection mechanism and software content and features are very different from our software.

8.3 BACKGROUND

In this section, we describe some preliminary concepts fundamental for trajectory generation from raw videos. Figure 8.1 demonstrates the video processing pipeline. Using computer vision techniques, we process the raw fisheye videos and detect and track objects at an intersection, such as cars and pedestrians. We cluster the generated trajectories as a post-processing step before visualization.

8.3.1 VIDEO ANALYSIS

Video analysis deals with the algorithms needed to process the raw videos to generate trajectories. Our software has been developed to process videos from fisheye cameras, but it can easily be extended to process videos from ordinary cameras. The benefit of using a fisheye camera is that it can capture the whole intersection in a wide panoramic and nonrectilinear image using its wide-angled fisheye lens. This allows a fisheye camera to capture the entire intersection in a single view. In some cases, where the intersection is large, two or more fisheye cameras are often used to get a complete picture.

Detecting and tracking objects using timestamped coordinates leads to the generation of object trajectories. For a typical intersection that is moderately busy, there are over 10,000 trajectories happening on a weekday. Thus, a visualization system helps monitor traffic behavior automatically without watching traffic videos. In our software, we automatically anonymize the information about the moving object by saving to our database only the location coordinates of objects and, in the cases of vehicles, their size and color.

As a result of video processing, a video is converted to a series of frames. The objects are detected and tracked across the frames in all of these frames. Video processing uses a temporal superpixel (supervoxel) method [61] to extract an accurate mask for object representation. These can be converted into trajectories representing traffic's spatial and temporal movement. A trajectory is a path traversed by a moving object represented as successive spatial coordinates and corresponding timestamps. The details of our video processing and analysis system are presented in detail in Ref. [73].

8.3.2 TRAJECTORY DATABASE

We use a relational database in the cloud to store the trajectories generated by our software and the signal phase and timing information obtained from high-resolution controller logs. Figure 8.2 presents the key attributes found in the database. The attributes of TrackInfo represent the properties of the trajectories we consider for analysis. These are (1) frame_id, which identifies the current video frame, (2) track_id, which determines a trajectory, (3) x and y, the coordinates of the object location, (4) w and h, the width and height, respectively, of the bounding box enclosing the object, (5) intersection_id, which identifies the intersection, and (6) date and time, which represent the timestamp. Correspondingly, the attributes for signal data include (1) intersection_id, (2) timestamp, (3) the current state of the signals encoded in a hexadecimal format, (4) the cycle number (relative to the cycle number of the first observation), and (5) camera_id, which is an additional attribute that depends on the intersection.

8.3.3 TRAJECTORY PROCESSING

Given a video feed, one of the first tasks is to generate the trajectories. We summarize the steps involved in generating the trajectories here. Each trajectory is preprocessed to eliminate coordinates that appear outside the intersection scope. This step is helpful because trajectories farther away from the fisheye cameras tend to be noisy. After the preprocessing, we fuse the signal timing data with the tracks. An unsupervised clustering approach is used to cluster the trajectories as a final step.

8.3.4 FUSION WITH SIGNAL DATA

Because we focus on capturing trajectories through an intersection, we observe a definite pattern of movement brought about by the presence of through and turn lanes and the signaling system at an intersection. The advanced transportation controllers (ATCs) allow us to record the exact time for signal changes and detector on/off events ten times a second. We download the recorded data for the intersection for a given period.

The signal controller records the signal changes to an intersection's red, yellow, or green light. We store the current intersection state in which signals are red/yellow/green using a 24-bit binary number and its hexadecimal equivalent. This

```
mysql> select * from VideoTracks limit 5;
+-----+-----+-----+-----+-------+-------------------------+------+-----+
| fid | tid |  x  |  y  | class | timestamp               | iid  | cid |
+-----+-----+-----+-----+-------+-------------------------+------+-----+
|  12 |   2 | 537 | 217 |  car  | 2019-07-16 11:00:13.200 | 5225 |  6  |
|  13 |   2 | 538 | 217 |  car  | 2019-07-16 11:00:13.300 | 5225 |  6  |
|  14 |   2 | 539 | 217 |  car  | 2019-07-16 11:00:13.400 | 5225 |  6  |
|  15 |   2 | 540 | 217 |  car  | 2019-07-16 11:00:13.500 | 5225 |  6  |
|  16 |   2 | 541 | 217 |  car  | 2019-07-16 11:00:13.600 | 5225 |  6  |
+-----+-----+-----+-----+-------+-------------------------+------+-----+
5 rows in set (0.04 sec)
```

```
mysql> select * from OnlineSPaT limit 5;
+-----+-----+-------------------------+----------+-------+
| iid | cid | timestamp               | hexphase | cycle |
+-----+-----+-------------------------+----------+-------+
| 659 |  5  | 2019-11-23 10:45:10.000 | 4400bb   |     1 |
| 659 |  5  | 2019-11-23 10:45:10.000 | 4400bb   |     1 |
| 659 |  5  | 2019-11-23 10:45:17.000 | 0044bb   |     1 |
| 659 |  5  | 2019-11-23 10:45:32.000 | 0000ff   |     1 |
| 659 |  5  | 2019-11-23 10:45:37.900 | 0044bb   |     1 |
+-----+-----+-------------------------+----------+-------+
5 rows in set (0.04 sec)
```

Figure 8.2 MySQL databases of trajectories generated by video processing (*left*) and the signal information (*right*) generated from ATSPM

encoding works out as follows: We divide the 24 bits into three sets of 8 bits each. The first set of 8 bits is reserved for recording the green status on each phase. The green status may be ON(1) or OFF(0) for each phase 1 through 8. Similarly, the second and third sets of 8 bits are reserved for recording yellow and red status, respectively. This encoding is explained in the following example. Suppose phases 4 and 8 are green, and every other signal is red. Then the 24-bit signal state is *0001 0001 0000 0000 1110 1110*. The equivalent hexadecimal representation is *1100ee*.

So, now we have two sets of data – video and signal phasing data. Our next step is to fuse the data and annotate each trajectory coordinate with the signal state when it happens. This is possible because the video processor annotates each trajectory coordinate with the timestamp when that coordinate is traversed. Thus, upon starting to process a new video, the first thing is synchronizing the video timestamps with the controller timestamps. The video processor's timestamps are usually a few seconds later than those on the controller logs. The offset results are due to the time difference between starting the command and receiving the video stream.

8.3.5 CLUSTERING TRAJECTORIES

For clustering the trajectories, we need a distance measure that is appropriate for the trajectories. We developed a distance measure based on FastDTW [171] to compute

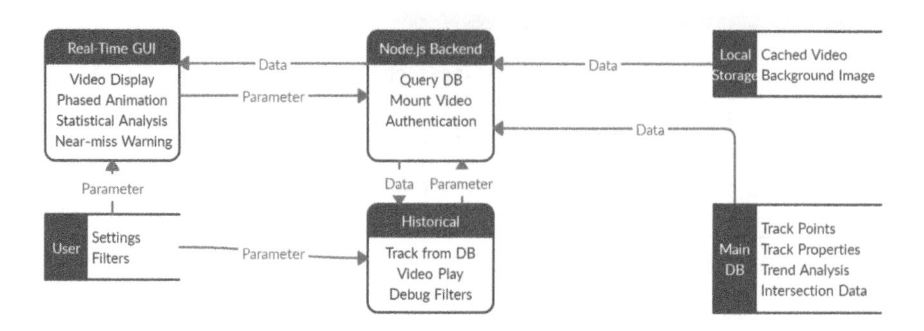

Figure 8.3 The data flow diagram of the whole process

a list of coordinate pairs from the two trajectories that align with each other while considering the different trajectory speeds. FastDTW approximates the dynamic time warping algorithm to find the optimal alignment between two-time series with near-linear time and space complexity. To compute the distance between the trajectories, we use the coordinate alignment pairs from DTW to divide the space between the trajectories. Then, we calculate the sum of the area of the triangles and divide it by the average length of the two trajectories to determine the average distance between the trajectories.

Following the computation of the distance between every pair of trajectories, we apply spectral clustering to obtain the trajectory clusters.

8.4 VISUALIZATION

Our visualization software is divided into two parts – the frontend and the backend. The frontend is written in Vue.js, one of the most popular progressive frameworks for building user interfaces. The frontend has two modes – the streaming mode and the historical mode. The streaming mode loads the video and corresponding starting timestamp from the backend and then uses a cache to query the database every 5 seconds for three operations – track animation, the statistics related to the tracks, and the display of tracks by phases. In the historical mode, on the other hand, the software reads the database based on filters that the user can select. The backend is written in Node.js, whose primary purpose is to query the database. The backend loads the Ajax post from the frontend and then queries the database using the received data. The backend also prepares the video file or stream to be displayed in the frontend. We now describe the capabilities of our software in more detail, first for the streaming mode visualization in Section 8.4.1 and then for historical analysis in Section 8.4.2.

A data flow diagram (DFD) describes the software's overall structure. Figure 8.3 shows how the data goes through the pipeline and finally appears in the interface.

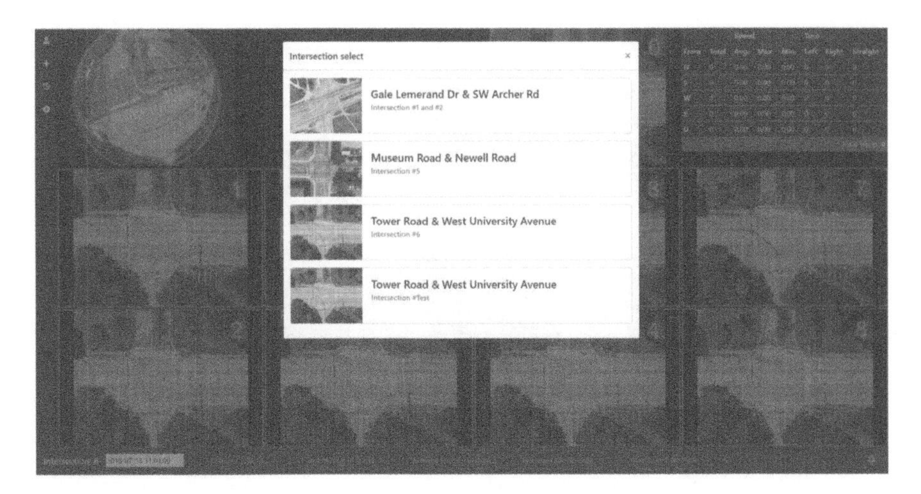

Figure 8.4 The selection of the current intersection can be opened by clicking on the button on the sidebar

8.4.1 STREAMING VISUALIZATION

After connecting to the tool, the user logs in and chooses the intersection of interest. The intersection selection step is shown in Figure 8.4. The tool supports the analysis of multiple intersections simultaneously. The streaming mode page, as shown in Figure 8.5, is the first page the user sees after logging into the system and choosing the intersection of interest. The various components of this page are described below.

8.4.1.1 Playing Video

The fisheye video of the intersection, as captured by a fisheye camera, is streamed or played on the top-left corner of the web page. An animation of the object trajectories is created by representing the objects as moving dots at the intersection. This animation is displayed beside the video in an unwarped space and in a time-synchronized manner for the actual video.

For streaming mode accessibility of the data, we use database caches or put the data in memory where appropriate. In other words, we use a buffer to store all the data we get from the database. We set a timer to read the max timestamp stored in the buffer while comparing it with the current timestamp obtained from video play. Once the max timestamp of the buffer is less than 5 seconds later than the current one, the program queries the database and reads 5 seconds of data after the max timestamp. Thus, 5–10 seconds of future data is stored in the buffer every time. As the buffer size increases, the timer evicts the last hour of data when the total data reaches 3 hours. A pointer for the start point of the tracks updates every animation frame for a faster query of animated tracks.

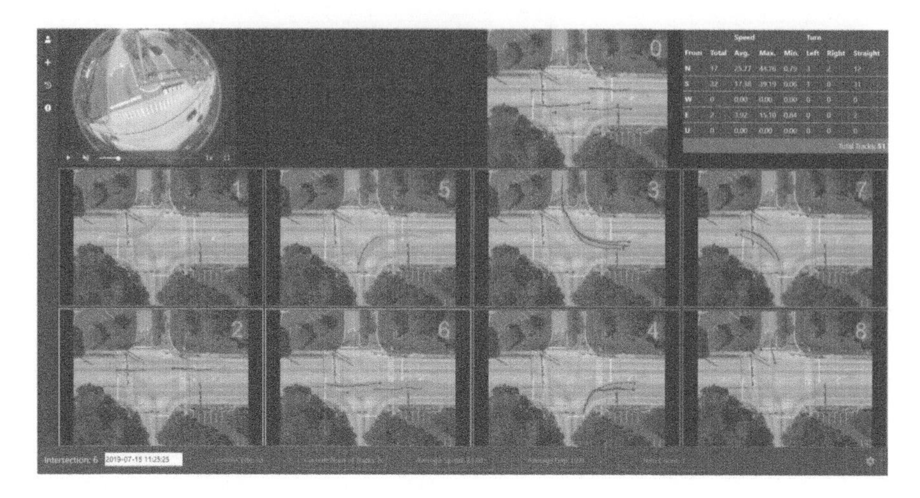

Figure 8.5 Interface for processing streamed tracks. The video from the fisheye camera is at the top-left corner. This is followed by an unwarped image of the intersection and the trajectories represented by a moving dot. The statistics of the trajectories on the intersection are captured in the table on the top-right side. The trajectories are partitioned into phases and displayed in the eight windows. The setting options and average statistics about the trajectories are presented at the bottom

8.4.1.2 Displaying Phases

There is a window for each of the eight phases of traffic movement, as shown in Figure 8.5. The eight windows are tiled and numbered according to the phase they represent. An ongoing trajectory is assigned a phase by first performing matches with the existing centroids or cluster centers. The trajectory phase is set to the phase of the matching centroid.

After the phase is assigned, the trajectory is displayed in the respective phase window. Separating the trajectories by phases makes them easy to track and helps us count the number of vehicles and their classes for each movement. The software supports an overlay on the phase windows that displays the number of vehicles in the through and turn lanes for that movement for a given time window, shown in Figure 8.6. The user may enable the overlay by clicking on the settings and then toggling the "Enable Overlay" switch.

It is possible to zoom in on only one phase by clicking on that phase. The result is shown in Figure 8.7, where phase 6 was selected by the user,

8.4.1.3 Displaying Signal Data

Our software integrates the trajectories with signal data to place the trajectories in the context of the current signaling state of the intersection. This signal data integration enables us to detect anomalous traffic behavior at the intersection, such as vehicles

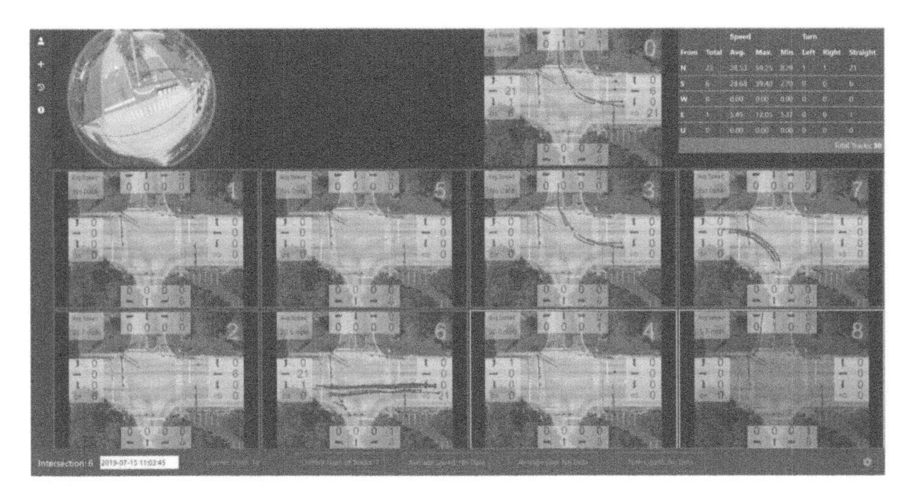

Figure 8.6 The overlay feature of the tool may be used to display the vehicle egress and ingress counts for the different phases of movement

moving in the yellow signals or those jumping the red. The signal data information in the GUI is displayed by framing each window with a box of the same color as that of the signal for that phase. Phases 2 and 6 are being served in Figure 8.5; hence, these phase windows are framed by green boxes in the figure.

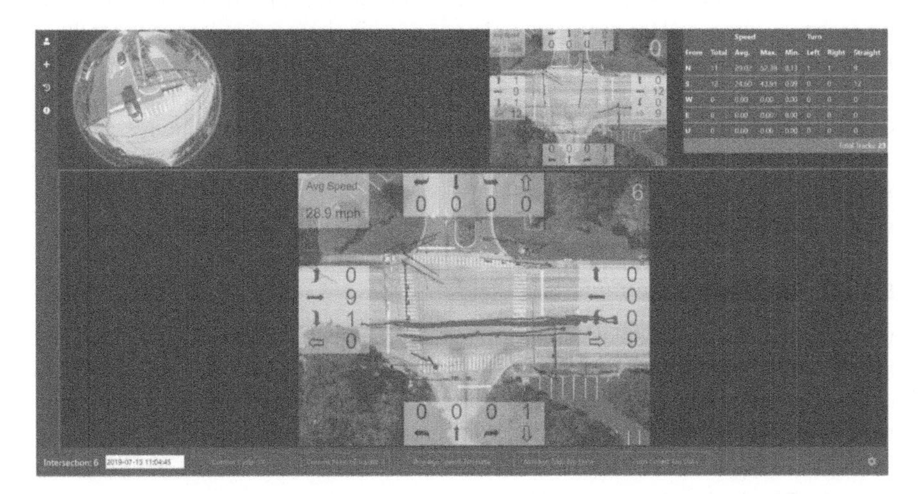

Figure 8.7 A particular phase was zoomed in by clicking on that phase from the main page. The user can toggle to the main page by clicking the zoomed-in phase

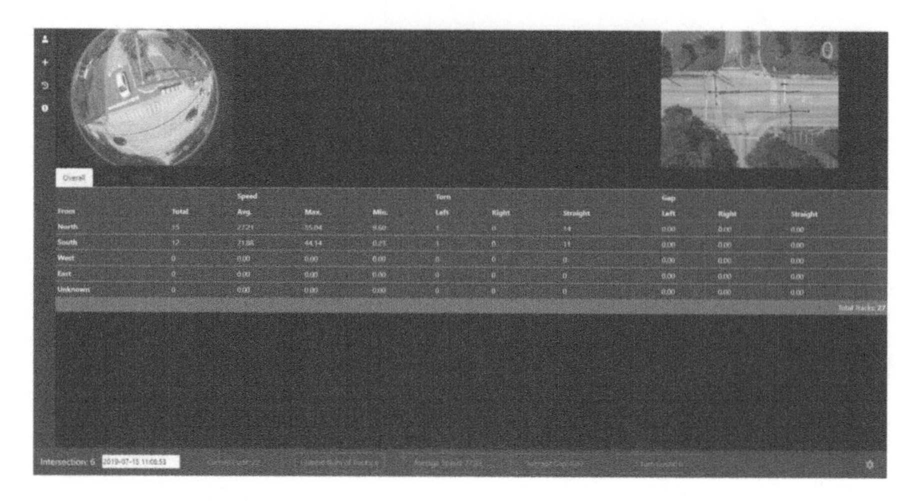

Figure 8.8 Clicking on the statistics table on the main page toggles the display to more statistical details. These are organized into three tabs: Overall, Cycle, and Charts. The overall statistics for speed, gap, and turn movement count are here

8.4.1.4 Displaying Statistics

The statistics for the various trajectories are available in the form of charts and tables. The table appears on the top-right corner of the streaming page of Figure 8.5. The table shows the most common statistics, such as the number of vehicles or pedestrians for each movement and the average, minimum, and maximum speeds in mph for those movements. If the user is interested in more details, such as the average gap or the response time, the table may be clicked to bring up the details. The further details are organized into three tabs: Overall, Cycle, and Charts, as shown in Figures 8.8, 8.9, and 8.10, respectively. The Overall tab gives the overall observations, as the name suggests. The Cycle tab aggregates the data on a cycle-by-cycle basis. The Chart tab, on the other hand, shows the overall trend of the characteristics over time.

8.4.1.5 Displaying Multi-camera Views

Our visualization tool displays a merged image from two or more cameras typically installed in a large intersection. Figure 8.11 shows one such intersection, where two cameras are installed in the opposite corners of a large intersection. The animation merges the views from the two cameras and creates a single image.

8.4.1.6 Displaying Near-miss Events

Our tool automatically picks up the near-miss and crash events and reports separately. Such incidents are usually rare, but Figure 8.12 represents a hypothetical situation to demonstrate the feature.

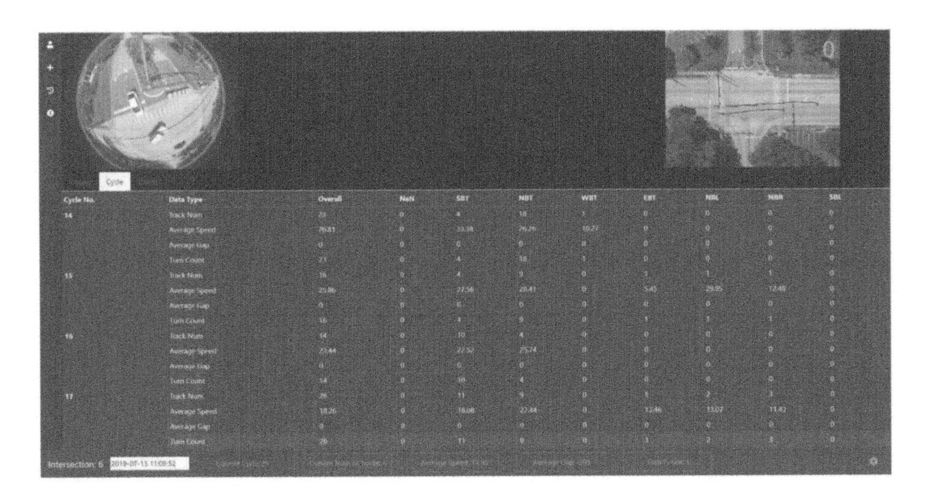

Figure 8.9 The detailed statistical data cycle tab shows the statistics data aggregated by signal cycles and displayed by cycle counts

8.4.1.7 Displaying Track Information

It is possible to get the track ids of the tracks currently being displayed by clicking on the "Current Num Of Tracks," which is shown in Figure 8.13. The user can view the detailed information on that path by choosing a track. These include the speed, cluster, and phase of the path.

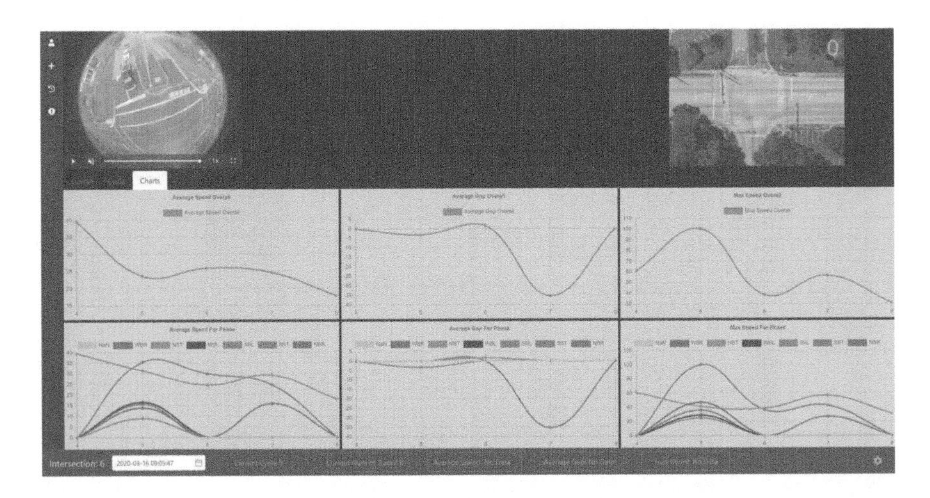

Figure 8.10 When clicked, the charts tab shows the speed and gap statistics visually as charts

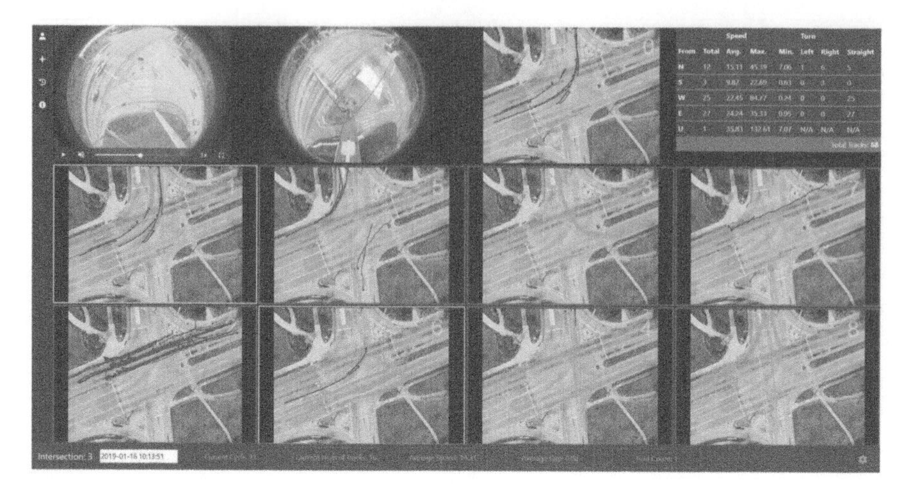

Figure 8.11 The images from an intersection with two cameras. The cameras are installed in opposite corners of the intersection, and the raw video from each camera appears in the top left-corner

8.4.1.8 Other Settings

In addition to the settings of the features described above, there are other valuable features, such as coloring trajectories by object classes or selecting the time frame for displaying tracks, such as those that happened in the last x seconds. Figure 8.14 shows the entire selection collection available to the user upon clicking the settings.

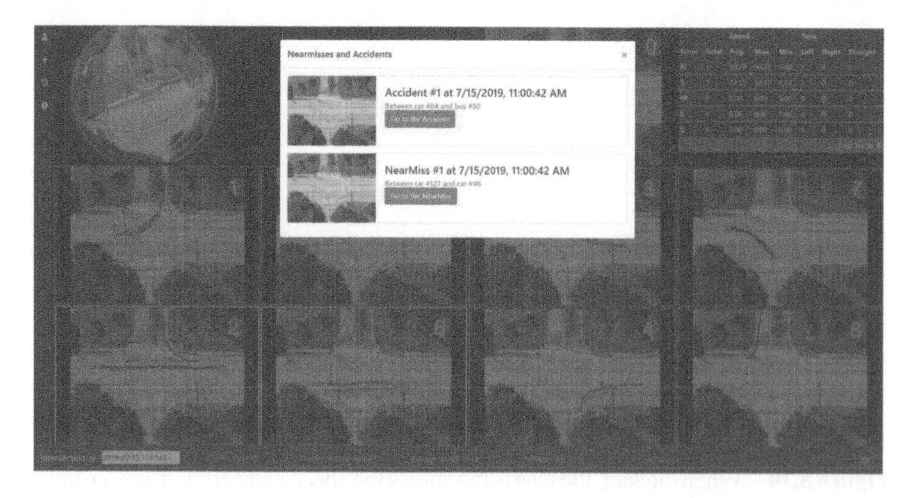

Figure 8.12 The interface for displaying near-miss and crash events in streaming mode. The two events described here are hypothetical and are used to test the interface

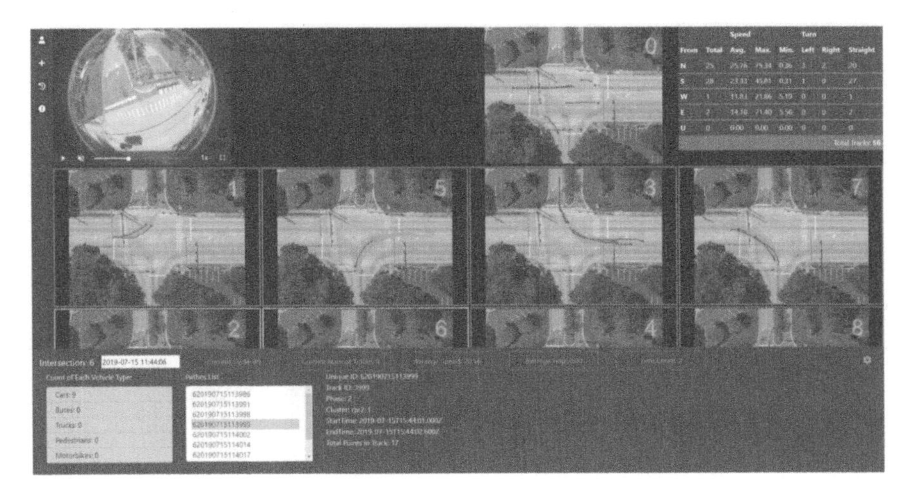

Figure 8.13 The charts tab, when clicked, shows the speed and gap statistics visually as tables

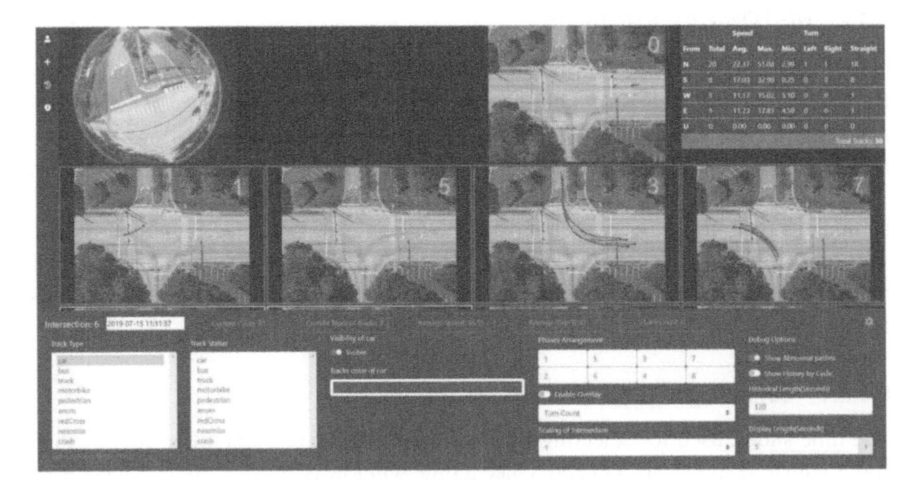

Figure 8.14 The settings menu. It offers several options to the user for enhanced display. It contains the option to enable overlay, set the color for trajectories based on class type, and set the window to display historical data within the streaming mode. The latter is usually set to about 2 minutes to capture all trajectories in an ongoing cycle.

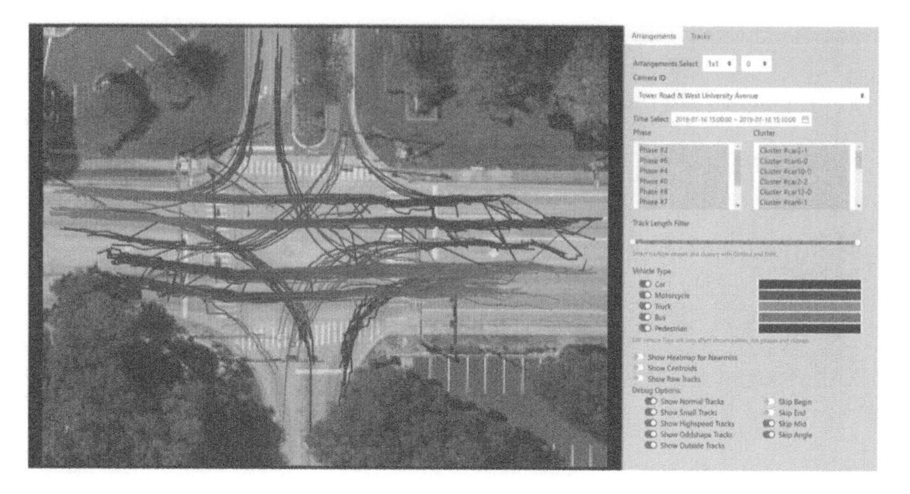

Figure 8.15 The display for historical analysis. It provides several options for filtering a large collection of trajectories. The primary filter is based on the time range. All tracks that happened for 10 minutes starting on Monday, July 15, 2019, 15:00:12 GMT, are displayed here

8.4.2 HISTORICAL ANALYSIS

Our visualization software may be used for mining the trajectories created in the past from traffic videos. In this section, we describe the features supported for historical analysis. Figure 8.15 shows the landing page for historical analysis. It allows the users to filter the trajectories based on time of occurrence, phase, and cluster.

Further, the user may view the cluster centroids of the dominating clusters, the anomalous trajectories, or even filter out the trajectories impacted by noises. It also supports the side-by-side display of trajectories for different combinations of time of the day, the day of the week, or classes of objects. We present some of the possibilities in this section.

8.4.2.1 Selecting a Time Window for Trajectories

The time slider, used for selecting a time window, is the primary filter. It is a range slider with a flexible start and end. Once a time window is chosen, all the trajectories occurring in that time window on the selected intersection are obtained using a database query and displayed on the screen. The trajectories in Figure 8.15 show such an example, where the time window length is 10 minutes starting Monday, July 15, 2019, 15:00:12 GMT.

8.4.2.2 Selecting Phases and Clusters

All trajectories in a given time window may be further partitioned based on their movement and cluster. Figure 8.16 shows all the eight phases the trajectories are

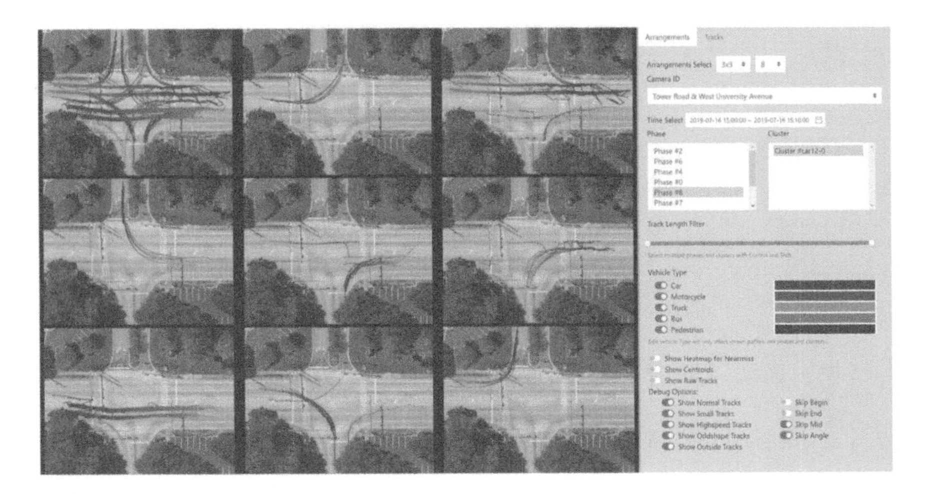

Figure 8.16 The tracks in the top-left corner are partitioned into different phases. Going row wise from top to bottom and from left to right of each row, the first window on the top-left corner shows all the tracks in the previously chosen time range. Then the tracks for phases 1, 2, 3, 4, 5, 6, 7, and 8 are displayed on the respective windows

partitioned into for the trajectories belonging to the time window chosen in Figure 8.15. The top-left window shows all the trajectories, whereas the other windows show the trajectories corresponding to the eight vehicle phases described in Ref. [69].

Figure 8.17 showcases a selection combination for phases and clusters. The trajectories in phase 6 here are partitioned into three clusters. Two of these, namely, "car6-0" and "car6-1", correspond to the trajectories on the two lanes, respectively, while "car11-0" corresponds to the right-turn trajectories. These three clusters are shown in the three remaining windows. The remaining cluster, "small," corresponds to the tiny tracks created due to processing errors and are not shown in this figure. Figures 8.16 and 8.17 demonstrate our software's powerful feature, enabling side-by-side comparison of sets of trajectories, where each set may be chosen independently using any criteria.

The "Phase" window often has a phase 0, simply a phase to which the trajectories are assigned when the software cannot determine their movement direction. This happens for anomalous trajectories and trajectories resulting from a processing error.

8.4.2.3 Cluster Centers

The visualization tool may display the cluster centers used for the online cluster assignment. This is useful in sampling the major movements at an intersection. It also comes in handy in debugging mismatched alignment and stitching small tracks. Figure 8.18 shows the cluster centers corresponding to the trajectories of the 10 minutes duration chosen earlier.

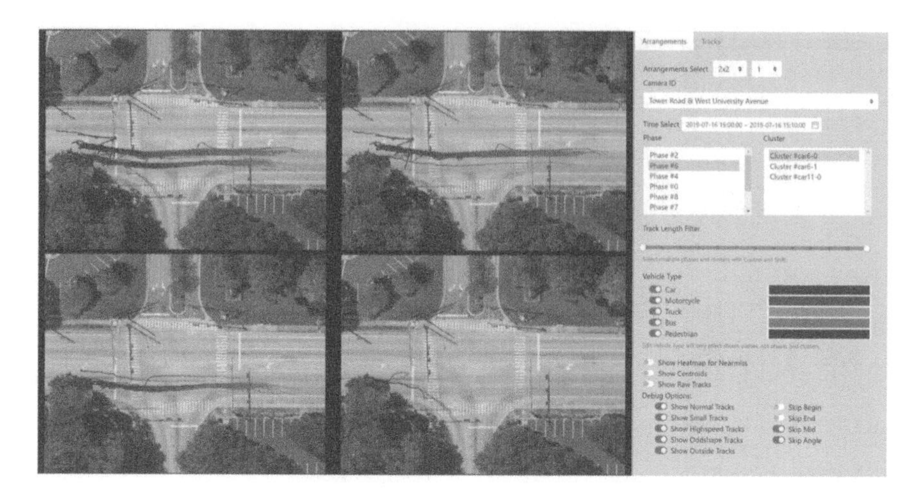

Figure 8.17 A combination of phases and clusters was used to display all trajectories in phase 6 and their respective clusters. These clusters automatically partition the trajectories based on their lane of occurrence and between the through and right-turn trajectories

8.4.2.4 Trajectories by Object Class

It is possible to filter the trajectories by the object classes, such as cars, buses, trucks, and pedestrians, which makes it easier to generate a count of each type of object for a given time window. Figure 8.19 shows an example of this.

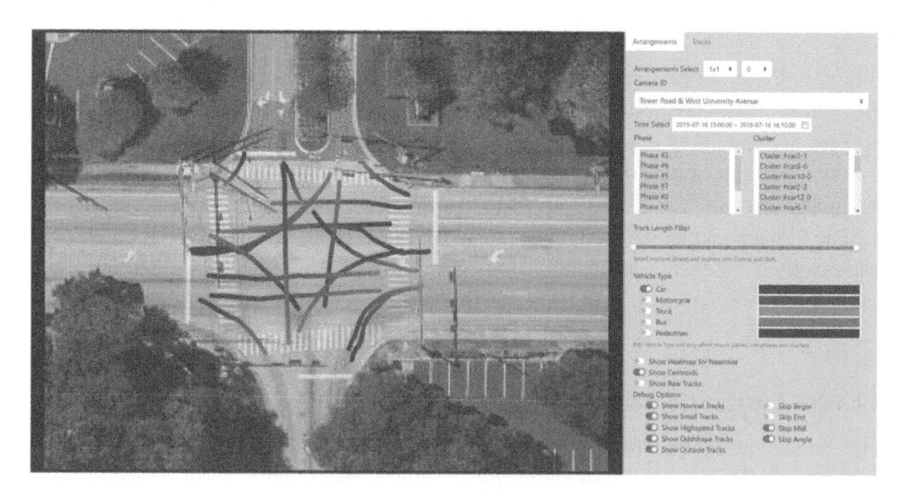

Figure 8.18 The centroids or the cluster centers of the trajectories

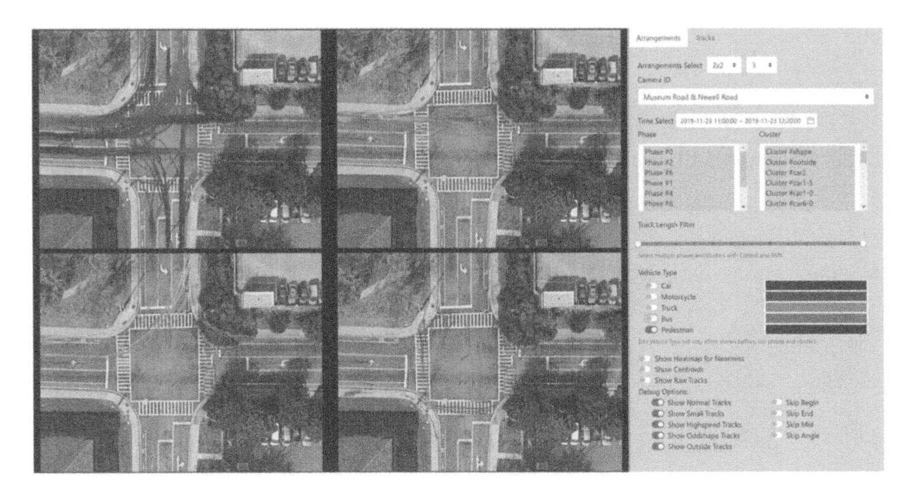

Figure 8.19 The trajectories are partitioned by the object classes. The top-left window shows all the trajectories made by cars, the top-right window shows the trajectories made by motorcycles, the bottom-left window shows trajectories generated by buses, and, finally, the bottom-right window shows trajectories generated by pedestrians

8.4.2.5 Heatmap for Near-misses

Figure 8.20 shows how heatmaps would appear to highlight unsafe areas of the intersection that have had more near-misses. The heatmap in this figure is based on a hypothetical dataset.

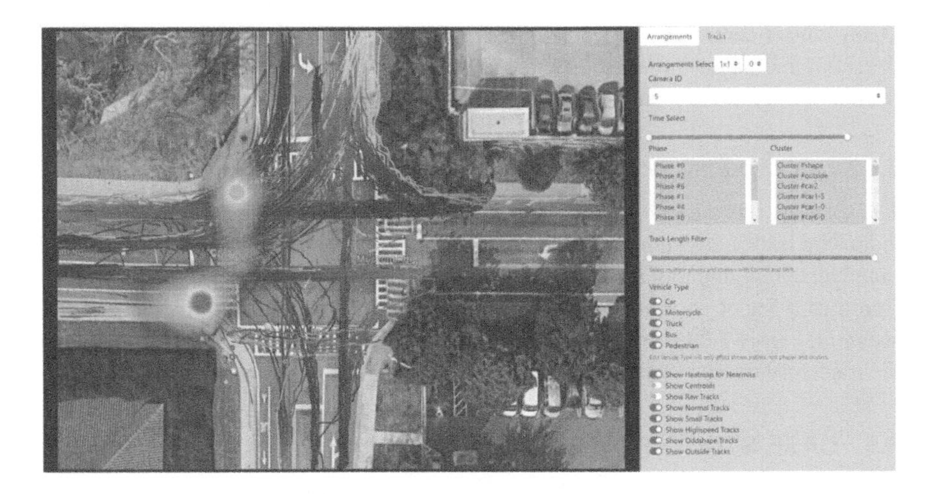

Figure 8.20 A heatmap based on hypothetical near-miss and crash events

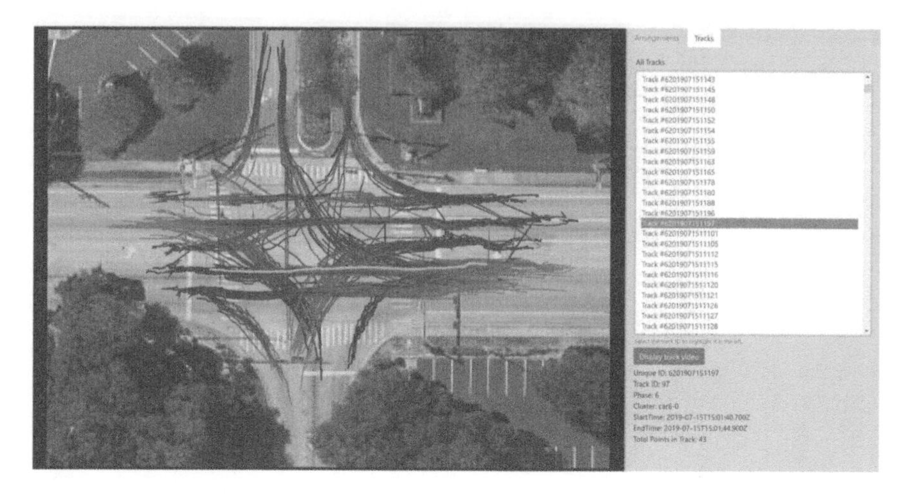

Figure 8.21 An individual trajectory selected by clicking on a track number from the Tracks drop-down

8.4.2.6 Individual Tracks

So far, we have focused on the display of sets of trajectories. Viewing the individual tracks by choosing a track from the Tracks tab is also possible. Figure 8.21 shows an individual highlighted trajectory. The highlighted trajectory's details, such as start time, end time, track ID, phase, and object class, are shown at the bottom-right corner.

8.4.2.7 Anomalous Tracks

A track may be anomalous for several reasons, such as when its shape does not conform to any cluster centroids or occurs while the ongoing signal is red. Figure 8.22 shows the set of anomalous trajectories with a trajectory we have highlighted that turns left from a through lane, even though there is a left-turn lane at this intersection.

8.4.2.8 Video Play for Selected Track

The software lets the user verify any trajectory in the historical domain by playing its video clip if the corresponding video is available. This feature makes our software an excellent debugging tool for developers because it closes the gap between a trajectory stored as a set of coordinates and the same trajectory as it traversed in a real-life scenario. Figure 8.23 shows a snapshot of the video being played that corresponds to the trajectory chosen in Figure 8.21.

8.5 CASE STUDY: TRAFFIC TREND ANALYSIS

This section aims to demonstrate how our software can effectively study traffic trends without observing hours of videos. Using the historical analysis interface, we can analyze trajectories from the same time interval over multiple days and discover trends.

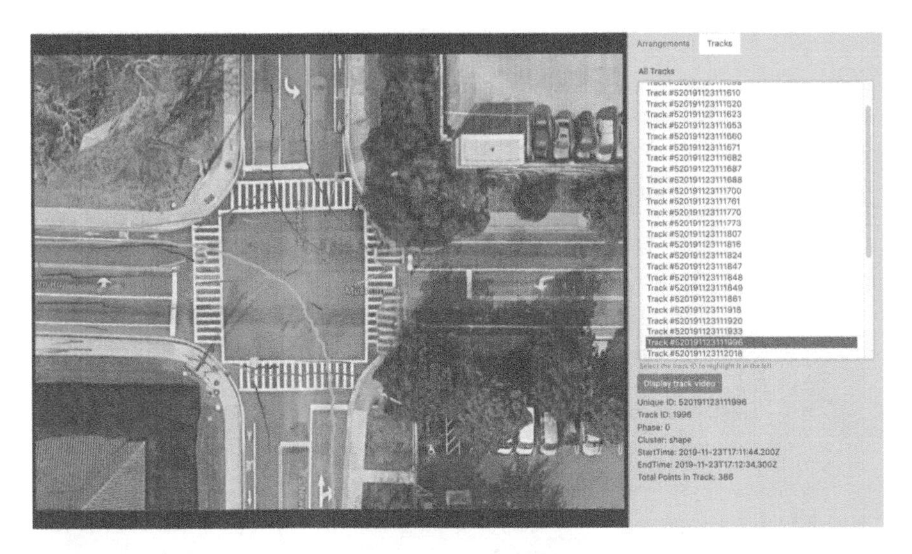

Figure 8.22 All trajectories that are identified as anomalous. The one highlighted takes a left turn from a through lane

Trajectories from multiple days may be loaded using the tiling feature described earlier. Figure 8.24 demonstrates the time of day and day of the week trends for the main clusters of car trajectories. The tiles' left, middle, and right columns are from

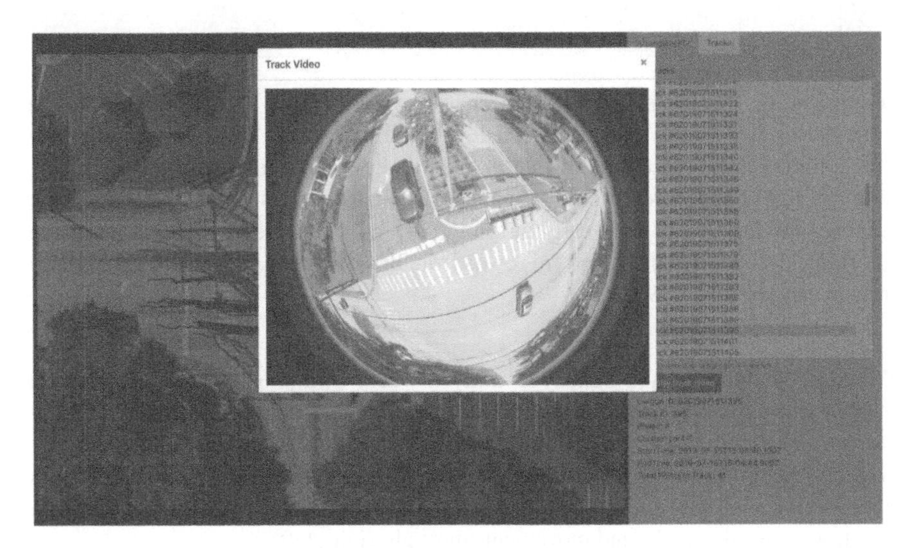

Figure 8.23 A video clip corresponding to the track selected in Figure 8.21 is being played

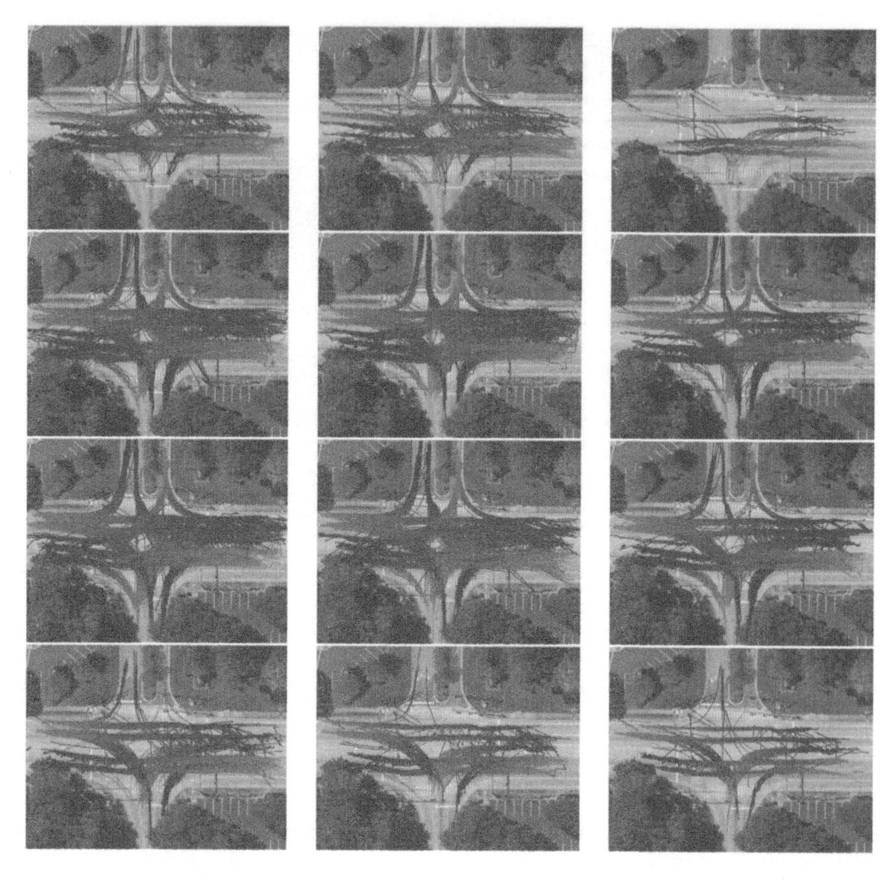

Figure 8.24 A case study for trend analysis. Each of the 12 tiles represents the trajectories over 2-hour intervals over a day for multiple days a week. In particular, the left column of tiles represents a Monday, the middle column represents a Wednesday, and the right column represents a Sunday. Each row of tiles is for a different time of day. The top row is from early morning (7–9 a.m.), the second row is from late morning (11 a.m.–1 p.m.), the third row is from the afternoon (3–5 p.m.), and the last row is late evening (7–9 p.m.)

Monday, Wednesday, and Sunday of a given week. On the other hand, the tiles on each row are from different intervals. The first row from the top shows data for early morning (7–9 a.m.), the second row of tiles shows data from mid-morning (11 a.m.–1 p.m.), the third row for the afternoon (3–5 p.m.), and the last row shows data from the evening (7–9 p.m.). Figure 8.25 gives the exact count of vehicles per direction (east, west, north, south) and movement (through, right, left).

We present some observations below based on these comparative vehicle counts by the time of day and week:

Time	7a.m. – 9a.m.			11a.m. – 1p.m.			3p.m. – 5p.m.			7p.m. – 9p.m.		
Day	Mon	Wed	Sun	Mon	Wed	Sun	Mon	Wed	Sun	Mon	Wed	Sun
EBL	20	87	0	151	134	13	129	95	9	12	0	0
EBR	0	2	0	159	51	0	129	111	4	0	0	0
EBT	0	8	0	6	4	2	13	2	2	0	0	0
NBL	144	218	0	63	51	3	27	16	0	0	0	0
NBR	110	99	22	129	67	146	182	138	140	114	79	14
NBT	1103	2102	446	1302	1146	1421	1219	1085	1077	761	643	560
SBL	50	106	43	199	134	128	194	134	156	135	145	145
SBR	81	93	0	115	125	0	52	51	14	0	0	52
SBT	380	749	199	1167	1172	1043	1530	1257	1141	1084	691	579
WBL	58	92	58	69	130	140	168	80	202	159	100	73
WBR	156	179	42	297	270	59	227	158	208	130	21	72
WBT	0	4	0	8	0	0	3	0	0	0	0	0
Total	2102	3739	810	3665	3284	2955	3873	3127	2953	2447	1679	1495

EBL	East Bound Left
EBR	East Bound Right
EBT	East Bound Through
NBL	North Bound Left
NBR	North Bound Right
NBT	North Bound Through
SBL	South Bound Left
SBR	South Bound Right
SBT	South Bound Through
WBL	West Bound Left
WBR	West Bound Right
WBT	West Bound Through

Figure 8.25 Turning and through movement counts for various periods in a single day

1. As expected, there is thin traffic on Sunday, especially during the early morning and late evening hours, compared to weekdays. It is easy to quantify this drop in vehicles as 60% and 40% fewer vehicles during the intervals 7–9 a.m. and 7–9 p.m., respectively, on a Sunday compared to the same time interval on Monday.
2. Similarly, if we consider the weekdays, then Monday is slightly different. The total volume of cars during the early morning hours is lower than that of Wednesday by about 50%. The volume of traffic increases on a Monday. About 40% more cars are there in the late evening on Monday than on Wednesday. This indicates a late start and a late end on Monday, the first day of the week.
3. For this particular intersection, we see a peak in northbound vehicles during the morning and an afternoon peak in southbound cars. This indicates that the significant residential areas nearby are toward the south.

8.6 DISCUSSION

In this chapter, we presented a novel software we developed for performing visual analytics. The software is designed to serve the broader user community of traffic practitioners in analyzing the efficiency and safety of an intersection. The software runs video analysis and generates the spatial locations and timestamps of all the objects passing through an intersection along with their object type.

As described in the chapter, our software has two modes: a streaming and a historical mode. These have their own merits and are helpful to a traffic engineer who can quickly analyze the trajectories to better understand traffic behavior at an intersection instead of watching long video sequences. For example, our software can diagnose red-light violations or high traffic in a turning bay. The traffic engineers can test appropriate remedies by performing before-and-after studies, and the problems may be reduced or eliminated.

The software is intuitive and straightforward to use, and it could help study trends in traffic patterns for different times of the day and the days of the week.

9 Conclusion

The implementation of computer vision and machine learning in intelligent transportation systems has excellent potential for various applications. However, challenges have arisen when deploying these applications, such as real-time analysis limitations and difficulty accommodating cameras with varying specifications. To address these issues, we have developed a range of video processing techniques and real-time multi-sensor-based frameworks specifically designed for both overhead and fisheye intersection videos.

In Chapter 2, we introduced a novel two-stream convolutional network architecture capable of real-time detection, tracking, and near-miss detection for vehicles in traffic video data. This architecture comprises two networks – a spatial and a temporal stream network. The spatial stream network detects individual vehicles and potential near-miss areas at the single-frame level using cutting-edge object detection methods to capture appearance features. Meanwhile, the temporal stream network utilizes motion features of the identified candidates to perform multiple object tracking, generating distinct trajectories for each tracked target.

Chapter 3 showcased a new unsupervised approach to detect near-misses in fisheye intersection video using a deep learning model combined with a camera calibration and spline-based mapping method. The method involved mapping road object coordinates from fisheye images onto an overhead map based on satellite imagery. Fisheye lens distortion and camera perspective distortion were corrected before mapping to achieve accurate distance and speed measurements. Adopting this integrated approach, the model performed real-time object recognition, multiple object tracking, and near-miss detection in fisheye video.

Chapter 4 details the development of algorithms that utilize video analysis and signal timing data to accurately detect and categorize events based on the phase and type of conflict in pedestrian–vehicle and vehicle–vehicle interactions. We also introduced a new surrogate safety measure, termed "severe event," quantified by metrics such as TTC or PET, as recorded in the event, deceleration, and speed. To filter the extensive set of conflicting interactions to a robust set of severe events, we devised an efficient multistage event filtering approach, followed by a multi-attribute decision tree approach.

Based on a limited number of intersections, our analysis discovered that the primary conflicts at intersections with heavy pedestrian use involved right-turning vehicles and pedestrians on the adjacent parallel crosswalk. We identified the specific right-turn directions contributing to this problem and the peak hours during the week when the issue occurred. Regarding vehicle–vehicle interactions, intersections with permissive left turns were found to have more common conflicts between left-turning vehicles and through vehicles. In contrast, intersections with protected left-only turns exhibited fewer merging and diverging conflicts, which are inherently less severe.

DOI: 10.1201/9781003431176-9

Chapter 5 demonstrated the utilization of trajectory, severe event, and signal phasing data to detect conflict and volume hotspots and analyze intersection performance. By thoroughly investigating and discussing the observed events at an intersection, we identified suitable countermeasures to address the issues. Furthermore, we evaluated the effectiveness of the chosen countermeasures in a transparent and reproducible manner.

Chapter 6 introduced a real-time trajectory prediction approach combined with aligned Google Maps information. Our method utilized a historical trajectory database to identify typical motion patterns that could guide future predictions. With this prior information, we could accurately predict trajectories based on the starting position, observation period, and prediction period. The efficacy and versatility of our approach were demonstrated through experimental results on three intersections. Through the application of this technology, we can proactively identify potential conflict situations and develop an early warning system for real-time vehicle–vehicle and pedestrian–vehicle conflicts.

Chapter 7 presented a new method and real-time system for estimating vehicle travel time using video processing across multiple intersections. Our experiments proved the proposed approach to be viable, effective, and scalable.

Chapter 8 focused on developing a novel visualization software for conducting visual analytics. The software was designed to serve traffic practitioners in analyzing the efficiency and safety of an intersection.

Our software offers two modes – a streaming mode and a historical mode. These modes have their unique benefits, and both are useful to traffic engineers who need to analyze trajectories quickly to better understand traffic behavior at an intersection. For example, our software can diagnose red light violations and heavy traffic in turning bays. Traffic engineers can test appropriate remedies by performing before-and-after studies, and our software can help to reduce or eliminate these issues.

Our software is user-friendly and easy to navigate, making it a valuable tool for analyzing traffic trends during various times and days of the week.

A Acknowledgments for Materials

We would like to thank Pan He, Siva Srinivasan, Bryce Grame, Emmanuel Posadas, Robert Schenck, Subhadipto Poddar, and Ahan Mishra for collaboration for portions of the work presented in this book.

We would like to thank Raj Ponnaluri, Trey Tillander, Darryll Dockstader, John Krause, Jeremy Dilmore, Tushar Patel, Ryan Casburn, Clay Packard, and Lily Elefteriadou for providing encouragement and support throughout this work. This work was supported in part by the Florida Department of Transportation (FDOT), FDOT District 5 and NSF SCC 1922782. The opinions, findings, and conclusions expressed in this book are those of the author(s) and not necessarily those of FDOT, the U.S. Department of Transportation and the National Science Foundation.

We would like to also acknowledge the following contributions toward the chapters in this monograph.

1. Chapter 1 used materials copyrighted by IEEE. Specifically, the definitive version appeared in the proceedings of IEEE International Conference on Connected and Autonomous Driving (MetroCAD), 2022 [172], and in the proceedings of IEEE International Conference on Connected and Autonomous Driving (MetroCAD), 2021 [173].
2. Chapter 2 used materials copyrighted by ACM. The definitive version appeared in *ACM Transactions on Spatial Algorithms and Systems*, 2020 [73].
3. Chapter 3 used materials from two sources. The definitive version of the clustering work appeared in the 6th International Conference on Vehicle Technology and Intelligent Transport Systems (VEHITS), 2020 [144], while the definitive version of the near-miss work appeared in the same conference, that is, the 6th International Conference on Vehicle Technology and Intelligent Transport Systems (VEHITS), 2020 [68].
4. Chapter 4 used materials copyrighted by IEEE. Specifically, the definitive version appeared in the proceedings of IEEE International Conference on Intelligent Transportation Systems (ITSC), 2022 [90].
5. Chapter 5 used materials published in MDPI Vehicles, 2022 [174].
6. Chapter 6 used materials copyrighted by IEEE. Specifically, the definitive version appeared in the proceedings of IEEE International Conference on Intelligent Transportation Systems (ITSC), 2021 [175].
7. Chapter 7 used materials published by MDPI. Specifically, the definitive version appeared in the *Journal of Imaging*, 2022 [176]

DOI: 10.1201/9781003431176-A

8. Chapter 8 utilizes materials copyrighted by Inderscience International Journal of Big Data Intelligence, 2020 [70]. Although Inderscience holds the copyright, they have granted us permission to reproduce the content in this monograph.

References

1. Amudapuram Mohan Rao and Kalaga Ramachandra Rao. Measuring urban traffic congestion: A review. *International Journal for Traffic & Transport Engineering*, 2(4): 286–305, 2012.
2. Jodi L Carson et al. *Best practices in traffic incident management*. Technical report, Federal Highway Administration, Office of Transportation, USA, 2010.
3. Glen Weisbrod, Don Vary, and George Treyz. Measuring economic costs of urban traffic congestion to business. *Transportation Research Record*, 1839(1):98–106, 2003.
4. Jonathan I. Levy, Jonathan J. Buonocore, and Katherine Von Stackelberg. Evaluation of the public health impacts of traffic congestion: A health risk assessment. *Environmental Health*, 9(1):65, 2010.
5. Kai Zhang and Stuart Batterman. Air pollution and health risks due to vehicle traffic. *Science of the Total Environment*, 450:307–316, 2013.
6. Muhammad Alam, Joaquim Ferreira, and Jose Fonseca. Introduction to intelligent transportation systems. *Intelligent Transportation Systems*, pp. 1–17, Springer, 2016.
7. Tawfik Borgi, Nesrine Zoghlami, and Mourad Abed. Big data for transport and logistics: A review. In *International Conference on Advanced Systems and Electric Technologies (IC ASET)*, pp. 44–49, 01 2017.
8. Robert Gordon. *Intelligent Transportation Systems*. Cham: Springer, 2016.
9. Benjamin Coifman, David Beymer, Philip McLauchlan, and Jitendra Malik. A real-time computer vision system for vehicle tracking and traffic surveillance. *Transportation Research Part C: Emerging Technologies*, 6(4):271–288, 1998.
10. Maria Valera and Sergio A Velastin. Intelligent distributed surveillance systems: A review. *IEE Proceedings: Vision, Image and Signal Processing*, 152(2):192–204, 2005.
11. Norbert Buch, Sergio A Velastin, and James Orwell. A review of computer vision techniques for the analysis of urban traffic. *IEEE Transactions on Intelligent Transportation Systems*, 12(3):920–939, 2011.
12. Shunsuke Kamijo, Yasuyuki Matsushita, Katsushi Ikeuchi, and Masao Sakauchi. Traffic monitoring and accident detection at intersections. *IEEE Transactions on Intelligent Transportation Systems*, 1(2):108–118, 2000.
13. Harini Veeraraghavan, Osama Masoud, and Nikolaos P. Papanikolopoulos. Computer vision algorithms for intersection monitoring. *IEEE Transactions on Intelligent Transportation Systems*, 4(2):78–89, 2003.
14. Pan He, Weilin Huang, Tong He, Qile Zhu, Yu Qiao, and Xiaolin Li. Single shot text detector with regional attention. In *Proceedings of the IEEE International Conference on Computer Vision*, pp. 3047–3055, 2017.
15. Alejandro Angel, Mark Hickman, Pitu Mirchandani, and Dmesh Chandnani. Methods of traffic data collection, using aerial video. In *Proceedings: The IEEE 5th International Conference on Intelligent Transportation Systems*, pp. 31–36, IEEE, 2002.
16. Giuseppe Salvo, Luigi Caruso, Alessandro Scordo, Giuseppe Guido, and Alessandro Vitale. Traffic data acquirement by unmanned aerial vehicle. *European Journal of Remote Sensing*, 50(1):343–351, 2017.

17. G. Scotti, L. Marcenaro, C. Coelho, F. Selvaggi, and C.S. Regazzoni. Dual camera intelligent sensor for high definition 360 degrees surveillance. *IEE Proceedings: Vision, Image and Signal Processing*, 152(2):250–257, 2005.
18. Ming-Liang Wang, Chi-Chang Huang, and Huei-Yung Lin. An intelligent surveillance system based on an omnidirectional vision sensor. In *2006 IEEE Conference on Cybernetics and Intelligent Systems*, pp. 1–6, IEEE, 2006.
19. Neil Hoose. IMPACTS: an image analysis tool for motorway surveillance. *Traffic Engineering & Control*, 33(3), 1992.
20. Dieter Koller, Kostas Daniilidis, and Hans-Hellmut Nagel. Model-based object tracking in monocular image sequences of road traffic scenes. *International Journal of Computer Vision*, 10(3):257–281, 1993.
21. Philip McLauchlan, David Beymer, Benn Coifman, and Jitendra Mali. A real-time computer vision system for measuring traffic parameters. In *Proceedings of IEEE Computer Society Conference on Computer Vision and Pattern Recognition*, pp. 495–501, IEEE, 1997.
22. Shaoqing Ren, Kaiming He, Ross Girshick, and Jian Sun. Faster R-CNN: Towards real-time object detection with region proposal networks. In *Advances in Neural Information Processing Systems*, pp. 91–99, 2015.
23. Ross Girshick, Jeff Donahue, Trevor Darrell, and Jitendra Malik. Region-based convolutional networks for accurate object detection and segmentation. *IEEE Transactions on Pattern Analysis and Machine Intelligence* 38(1):142–158, 2016.
24. Joseph Redmon, Santosh Divvala, Ross Girshick, and Ali Farhadi. You only look once: Unified, real-time object detection. In *Proceedings of the IEEE Conference on Computer Vision and Pattern Recognition*, pp. 779–788, 2016.
25. Zhi Tian, Weilin Huang, Tong He, Pan He, and Yu Qiao. Detecting text in natural image with connectionist text proposal network. In *European Conference on Computer Vision*, pp. 56–72, Springer, 2016.
26. Pan He, Aotian Wu, Xiaohui Huang, Jerry Scott, Anand Rangarajan, and Sanjay Ranka. Deep learning based geometric features for effective truck selection and classification from highway videos. *International IEEE Conference on Intelligent Transportation Systems (ITSC), 2019*. IEEE, 2019.
27. Pan He, Aotian Wu, Anand Rangarajan, and Sanjay Ranka. *Truck taxonomy and classification using video and weigh-in motion (WIM) technology final report*. Technical report, Final Research Report Prepared for Florida Department of Transportation, 2019.
28. Samy Sadeky, Ayoub Al-Hamadiy, Bernd Michaelisy, and Usama Sayed. Real-time automatic traffic accident recognition using HFG. In *2010 20th International Conference on Pattern Recognition*, pp. 3348–3351, IEEE, 2010.
29. Fu Jiansheng et al. Vision-based real-time traffic accident detection. In *Proceedings of the 11th World Congress on Intelligent Control and Automation*, pp. 1035–1038, IEEE, 2014.
30. Fan Jiang, Ying Wu, and Aggelos K. Katsaggelos. Abnormal event detection from surveillance video by dynamic hierarchical clustering. In *IEEE International Conference on Image Processing*, volume 5, pp. V–145, IEEE, 2007.
31. Stefan Hommes, Radu State, Andreas Zinnen, and Thomas Engel. Detection of abnormal behaviour in a surveillance environment using control charts. In *2011 8th IEEE International Conference on Advanced Video and Signal Based Surveillance (AVSS)*, pp. 113–118, IEEE, 2011.

32. Chang Liu, Guijin Wang, Wenxin Ning, Xinggang Lin, Liang Li, and Zhou Liu. Anomaly detection in surveillance video using motion direction statistics. In *IEEE International Conference on Image Processing*, pp. 717–720, IEEE, 2010.
33. Nacim Ihaddadene and Chabane Djeraba. Real-time crowd motion analysis. In *2008 19th International Conference on Pattern Recognition*, pp. 1–4, IEEE, 2008.
34. Shu Wang and Zhenjiang Miao. Anomaly detection in crowd scene. In *IEEE 10th International Conference on Signal Processing Proceedings*, pp. 1220–1223, IEEE, 2010.
35. Lijun Wang and Ming Dong. Real-time detection of abnormal crowd behavior using a matrix approximation-based approach. In *2012 19th IEEE International Conference on Image Processing*, pp. 2701–2704, IEEE, 2012.
36. Shuming Tang and Haijun Gao. Traffic-incident detection-algorithm based on nonparametric regression. *IEEE Transactions on Intelligent Transportation Systems*, 6(1):38–42, 2005.
37. Asim Karim and Hojjat Adeli. Incident detection algorithm using wavelet energy representation of traffic patterns. *Journal of Transportation Engineering*, 128(3):232–242, 2002.
38. Siyu Xia, Jian Xiong, Ying Liu, and Gang Li. Vision-based traffic accident detection using matrix approximation. In *2015 10th Asian Control Conference (ASCC)*, pp. 1–5, IEEE, 2015.
39. Lairong Chen, Yuan Cao, and Ronghua Ji. Automatic incident detection algorithm based on support vector machine. In *2010 Sixth International Conference on Natural Computation*, volumn 2, pp. 864–866, IEEE, 2010.
40. Yu Chen, Yuanlong Yu, and Ting Li. A vision based traffic accident detection method using extreme learning machine. In *2016 International Conference on Advanced Robotics and Mechatronics (ICARM)*, pp. 567–572, IEEE, 2016.
41. Iwao Ohe, Hironao Kawashima, Masahiro Kojima, and Yukihiro Kaneko. A method for automatic detection of traffic incidents using neural networks. In *Pacific Rim TransTech Conference. 1995 Vehicle Navigation and Information Systems Conference Proceedings. 6th International VNIS. A Ride into the Future*, pp. 231–235, IEEE, 1995.
42. Liu Yu, Lei Yu, Jianquan Wang, Yi Qi, and Huimin Wen. Back-propagation neural network for traffic incident detection based on fusion of loop detector and probe vehicle data. In *2008 Fourth International Conference on Natural Computation*, volume 3, pp. 116–120, IEEE, 2008.
43. Dipti Srinivasan, Xin Jin, and Ruey Long Cheu. Evaluation of adaptive neural network models for freeway incident detection. *IEEE Transactions on Intelligent Transportation Systems*, 5(1):1–11, 2004.
44. Samanwoy Ghosh-Dastidar and Hojjat Adeli. Wavelet-clustering-neural network model for freeway incident detection. *Computer-Aided Civil and Infrastructure Engineering*, 18(5):325–338, 2003.
45. Navneet Dalal and Bill Triggs. Histograms of oriented gradients for human detection. In *Proceedings of the IEEE Conference on Computer Vision and Pattern Recognition*. volumn 1, pp. 886–893, IEEE, 2005.
46. Alex Krizhevsky, Geoffrey Hinton, et al. *Learning Multiple Layers of Features from Tiny Images*. Technical report, University of Toronto, 2009.
47. Joseph Redmon and Ali Farhadi. Yolo9000: Better, faster, stronger. arXiv preprint, 2017.
48. Sergey Ioffe and Christian Szegedy. Batch normalization: Accelerating deep network training by reducing internal covariate shift. arXiv preprint arXiv:1502.03167, 2015.

49. Alex Bewley, Zongyuan Ge, Lionel Ott, Fabio Ramos, and Ben Upcroft. Simple online and realtime tracking. In *2016 IEEE International Conference on Image Processing (ICIP)*, pp. 3464–3468, IEEE, 2016.
50. Rudolph Emil Kalman. A new approach to linear filtering and prediction problems. *Journal of Basic Engineering*, 82(1):35–45, 1960.
51. Harold W Kuhn. The Hungarian method for the assignment problem. *Naval Research Logistics Quarterly*, 2(1–2):83–97, 1955.
52. Nicolai Wojke, Alex Bewley, and Dietrich Paulus. Simple online and realtime tracking with a deep association metric. In *2017 IEEE International Conference on Image Processing (ICIP)*, pp. 3645–3649, IEEE, 2017.
53. Wenhan Luo, Junliang Xing, Anton Milan, Xiaoqin Zhang, Wei Liu, Xiaowei Zhao, and Tae-Kyun Kim. Multiple object tracking: A literature review. arXiv preprint arXiv:1409.7618, 2014.
54. Xinchen Liu, Wu Liu, Huadong Ma, and Huiyuan Fu. Large-scale vehicle re-identification in urban surveillance videos. In *2016 IEEE International Conference on Multimedia and Expo (ICME)*, pp. 1–6, IEEE, 2016.
55. Nicolai Wojke and Alex Bewley. Deep cosine metric learning for person re-identification. In *2018 IEEE Winter Conference on Applications of Computer Vision (WACV)*, pp. 748–756, IEEE, 2018.
56. Tim Salimans and Durk P Kingma. Weight normalization: A simple reparameterization to accelerate training of deep neural networks. In *Advances in Neural Information Processing Systems*, pp. 901–909, 2016.
57. Radhakrishna Achanta, Appu Shaji, Kevin Smith, Aurelien Lucchi, Pascal Fua, and Sabine Süsstrunk. SLIC superpixels compared to state-of-the-art superpixel methods. *IEEE Transactions on Pattern Analysis and Machine Intelligence*, 34(11):2274–2282, 2012.
58. Pablo Arbelaez, Michael Maire, Charless Fowlkes, and Jitendra Malik. Contour detection and hierarchical image segmentation. *IEEE Transactions on Pattern Analysis and Machine Intelligence*, 33(5):898–916, 2011.
59. Varun Jampani, Deqing Sun, Ming-Yu Liu, Ming-Hsuan Yang, and Jan Kautz. Superpixel sampling networks. In *Proceedings of the European Conference on Computer Vision (ECCV)*, pp. 352–368, 2018.
60. Yupeng Yan, Xiaohui Huang, Anand Rangarajan, and Sanjay Ranka. Densely labeling large-scale satellite images with generative adversarial networks. In *2018 IEEE 16th International Conference on Dependable, Autonomic and Secure Computing, 16th International Conference on Pervasive Intelligence and Computing, 4th International Conference on Big Data Intelligence and Computing and Cyber Science and Technology Congress (DASC/PiCom/DataCom/CyberSciTech)*, pp. 927–934, IEEE, 2018.
61. Xiaohui Huang, Chengliang Yang, Sanjay Ranka, and Anand Rangarajan. Supervoxel-based segmentation of 3D imagery with optical flow integration for spatiotemporal processing. *IPSJ Transactions on Computer Vision and Applications*, 10(1):9, 2018.
62. Jean-Philippe Jodoin, Guillaume-Alexandre Bilodeau, and Nicolas Saunier. Urban tracker: Multiple object tracking in urban mixed traffic. In *IEEE Winter Conference on Applications of Computer Vision*, pp. 885–892, IEEE, 2014.
63. Stewart Jackson, Luis F Miranda-Moreno, Paul St-Aubin, and Nicolas Saunier. Flexible, mobile video camera system and open source video analysis software for road safety and behavioral analysis. *Transportation Research Record*, 2365(1):90–98, 2013.

64. Kilian Q. Weinberger and Lawrence K. Saul. Distance metric learning for large margin nearest neighbor classification. *Journal of Machine Learning Research*, 10:207–244, 2009.
65. Alexander Hermans, Lucas Beyer, and Bastian Leibe. In defense of the triplet loss for person reidentification. arXiv preprint arXiv:1703.07737, 2017.
66. Oren Rippel, Manohar Paluri, Piotr Dollar, and Lubomir Bourdev. Metric learning with adaptive density discrimination. arXiv preprint arXiv:1511.05939, 2015.
67. Fu-Hsiang Chan, Yu-Ting Chen, Yu Xiang, and Min Sun. Anticipating accidents in dashcam videos. In *Asian Conference on Computer Vision*, pp. 136–153, Springer, 2016.
68. Xiaohui Huang, Tania Banerjee, Ke Chen, Naga Venkata Sai Varanasi, Anand Rangarajan, and Sanjay Ranka. Machine learning based video processing for real-time near-miss detection. In *VEHITS*, pp. 169–179, 2020.
69. Federal Highway Administration, U.S. Department of Transportation. Traffic Signal Timing Manual. https://ops.fhwa.dot.gov/publications/fhwahop08024/chapter4.htm, 06 2008.
70. K. Chen, T. Banerjee, X. Huang, A. Rangarajan, and S. Ranka. A visual analytics system for processed videos from traffic intersections. In *6th International Conference on Vehicle Technology and Intelligent Transport Systems (VEHITS 2020)*, 2020.
71. Fred L. Bookstein. Principal warps: Thin-plate splines and the decomposition of deformations. *IEEE Transactions on Pattern Analysis and Machine Intelligence*, 11(6):567–585, 1989.
72. Haili Chui and Anand Rangarajan. A new point matching algorithm for non-rigid registration. *Computer Vision and Image Understanding* 89(2-3):114–141, 2003.
73. Xiaohui Huang, Pan He, Anand Rangarajan, and Sanjay Ranka. Intelligent intersection: Two-stream convolutional networks for real-time near-accident detection in traffic video. *ACM Transactions on Spatial Algorithms and Systems* 6(2), January 2020.
74. John C. Hayward. Near miss determination through use of a scale of danger. *Highway Research Record*, Issue 384:24–34, 1972.
75. Brian L Allen, B. T. Shin, and Peter J. Cooper. Analysis of traffic conflicts and collisions. *Transportation Research Record*, Issue 667:67–74, 1978.
76. S.M. Sohel Mahmud, Luis Ferreira, Md. Shamsul Hoque, and Ahmad Tavassoli. Application of proximal surrogate indicators for safety evaluation: A review of recent developments and research needs. *IATSS Research*, 41(4):153–163, 2017.
77. X. Shi, Y.D. Wong, M.Z.F. Li, and C. Chai. Key risk indicators for accident assessment conditioned on pre-crash vehicle trajectory. *Accident Analysis and Prevention*, 117:346–356, 2018.
78. Ashutosh Arun, Md Mazharul Haque, Ashish Bhaskar, Simon Washington, and Tarek Sayed. A systematic mapping review of surrogate safety assessment using traffic conflict techniques. *Accident Analysis and Prevention*, 153:106016, 2021.
79. Christer Hyden and L. Linderholm. The Swedish traffic-conflicts technique. *International Calibration Study of Traffic Conflict Techniques*: pp. 133–139, 1984.
80. A.R.A. van der Horst and A.R. van der Horst. *A Time-based Analysis of Road User Behaviour in Normal and Critical Encounters*. Institute for Perception TNO, 1990.
81. Tarek Sayed, Mohamed Zaki, and Jarvis Autey. Automated safety diagnosis of vehicle–bicycle interactions using computer vision analysis. *Safety Science* 59:163–172, 2013.
82. Jeffery Michael Archer. Indicators for traffic safety assessment and prediction and their application in micro-simulation modelling: A study of urban and suburban intersections. In *trb.org*, 2005.

83. Aliaksei Laureshyn and Andras Varhelyi. *The Swedish Traffic Conflict Technique: Observer's Manual*, 2018.
84. Corinna Cortes and Vladimir Vapnik. Support-vector networks. *Machine Learning*, 20(3):273–297, 1995.
85. L. Samara, P. St-Aubin, F. Loewenherz, N. Budnick, and L. Miranda-Moreno. *Video-based Network-wide Surrogate Safety Analysis to Support a Proactive Network Screening Using Connected Cameras: Case Study in the City of Bellevue (WA)*. Technical report, Transportation Research Board, 2021.
86. Alexey Bochkovskiy, Chien-Yao Wang, and Hong-Yuan Mark Liao. YOLOv4: Optimal Speed and Accuracy of Object Detection. 2020.
87. FDOT. Fatality Analysis Reporting System (FARS). `https://www.nhtsa.gov/research-data/fatality-analysis-reporting-system-fars`.
88. FDOT. Historical Item Averages Cost. `https://www.fdot.gov/docs/default-source/roadway/DS/13/IDx/17870.pdf`.
89. Haruo Takeda, Masami Yamasaki, Toshio Moriya, Tsuyoshi Minakawa, Fumiko Beniyama, and Takafumi Koike. A video-based virtual reality system. In *Proceedings of the ACM Symposium on Virtual Reality Software and Technology*, pp. 19–25, New York, NY, 1999, Association for Computing Machinery.
90. Tania Banerjee, Ke Chen, Alejandro Almaraz, Rahul Sengupta, Yashaswi Karnati, Bryce Grame, Emmanuel Posadas, Subhadipto Poddar, Robert Schenck, Jeremy Dilmore, Siva Srinivasan, Anand Rangarajan, and Sanjay Ranka. A modern intersection data analytics system for pedestrian and vehicular safety. In *2022 IEEE 25th International Conference on Intelligent Transportation Systems (ITSC)*, pp. 3117–3124, 2022.
91. FDOT. Traffic Conflict Techniques for Safety and Operations: Observer's manual. `https://www.fhwa.dot.gov/publications/research/safety/88027/88027.pdf`.
92. USDOT. Traffic Analysis Toolbox Volume VI: Definition, Interpretation, and Calculation of Traffic Analysis Tools Measures of Effectiveness. `https://ops.fhwa.dot.gov/publications/fhwahop08054/sect6.htm`.
93. Douglas Gettman and Larry Head. Surrogate safety measures from traffic simulation models. *Transportation Research Record*, 1840(1):104–115, 2003.
94. Chen Wang and Nikiforos Stamatiadis. Surrogate safety measure for simulation-based conflict study. *Transportation Research Record*, 2386(1):72–80, 2013.
95. Dominique Lord and Fred Mannering. The statistical analysis of crash-frequency data: A review and assessment of methodological alternatives. *Transportation Research Part A: Policy and Practice*, 44(5):291–305, 2010.
96. Mohamed Abdel-Aty and Anurag Pande. Crash data analysis: Collective vs. individual crash level approach. *Journal of Safety Research*, 38(5):581–587, 2007.
97. Xinguo Jiang, Guopeng Zhang, Wei Bai, and Wenbo Fan. Safety evaluation of signalized intersections with left-turn waiting area in China. *Accident Analysis and Prevention*, 95, 461–469, 2016.
98. F. H. Amundsen. Workshop on traffic conflicts. In *Proceedings: First Workshop on Traffic Conflicts Oslo 77. Norwegian Council for Scientific and Industrial Research*, 1977.
99. Katja Vogel. A comparison of headway and time to collision as safety indicators. *Accident Analysis and Prevention*, 35(3):427–433, 2003.

100. Lakshmi N. Peesapati, Michael P. Hunter, and Michael O. Rodgers. Evaluation of postencroachment time as surrogate for opposing left-turn crashes. *Transportation Research Record,* 2386(1):42–51, 2013.
101. G. Feng, S. G. Klauer, M.T. McGill, and T. Dingus. Evaluating the relationship between near-crashes and crashes: can near-crashes serve as a surrogate safety metric for crashes? *Transportation Research Board,* 811:382, Oct. 2010.
102. Carl Johnsson, Aliaksei Laureshyn, and Tim De Ceunynck. In search of surrogate safety indicators for vulnerable road users: A review of surrogate safety indicators. *Transport Reviews*, 38(6):765–785, 2018.
103. Ting Fu, Luis Miranda-Moreno, and Nicolas Saunier. A novel framework to evaluate pedestrian safety at non-signalized locations. *Accident Analysis and Prevention*, 111:23–33, 2018.
104. Wenqiang Chen, Tao Wang, Yongjie Wang, Qiong Li, Yueying Xu, and Yuchen Niu. Lane-based Distance–Velocity model for evaluating pedestrian–vehicle interaction at non-signalized locations. *Accident Analysis and Prevention*, 176:106810, 2022.
105. Di Yang, Kun Xie, Kaan Ozbay, and Hong Yang. Fusing crash data and surrogate safety measures for safety assessment: Development of a structural equation model with conditional autoregressive spatial effect and random parameters. *Accident Analysis and Prevention*, 152:105971, 2021.
106. Vittorio Astarita, Demetrio Carmine Festa, Vincenzo Pasquale Giofrè, and Giuseppe Guido. Surrogate safety measures from traffic simulation models a comparison of different models for intersection safety evaluation. *Transportation Research Procedia*, 37:219–226, 2019.
107. Mark Morando, Qingyun Tian, Long Truong, and Hai Vu. Studying the safety impact of autonomous vehicles using simulation-based surrogate safety measures. *Journal of Advanced Transportation* 2018, 2018.
108. Chen Wang, Yuanchang Xie, Helai Huang, and Pan Liu. A review of surrogate safety measures and their applications in connected and automated vehicles safety modeling. *Accident Analysis and Prevention*, 157:106157, 2021.
109. R.L. Anderson. Electromagnetic loop vehicle detectors. *IEEE Transactions on Vehicular Technology*, 19(1):23–30, 1970.
110. Yashaswi Karnati, Dhruv Mahajan, Anand Rangarajan, and Sanjay Ranka. Machine learning algorithms for traffic interruption detection. In *2020 Fifth International Conference on Fog and Mobile Edge Computing (FMEC)*, pp. 231–236, 2020.
111. Huijing Zhao, Jinshi Cui, Hongbin Zha, Kyoichiro Katabira, Xiaowei Shao, and Ryosuke Shibasaki. Monitoring an intersection using a network of laser scanners. In *2008 11th International IEEE Conference on Intelligent Transportation Systems*, pp. 428–433, 2008.
112. Bo Ling, Michael I. Zeifman, and David R.P. Gibson. Multiple pedestrian detection using IR LED stereo camera. In David P. Casasent, Ernest L. Hall, and Juha Roñing, editors, *Intelligent Robots and Computer Vision XXV: Algorithms, echniques, and Active Vision*. volume 6764, pp. 67640A. International Society for Optics and Photonics. SPIE, 2007.
113. Stefano Messelodi, Carla Modena, and Michele Zanin. A computer vision system for the detection and classification of vehicles at urban road intersections. *Pattern Analysis and Applications*, 8:17–31, 09 2005.
114. Ted Morris. Rapidly Deployable Low-Cost Traffic Data and Video Collection Device. https://www.cts.umn.edu/publications/report/rapidly-deployable-lowcosttraffic-data-and-video-collection-device, 2009.

115. Sokémi René Emmanuel Datondji, Yohan Dupuis, Peggy Subirats, and Pascal Vasseur. A survey of vision-based traffic monitoring of road intersections. *IEEE Transactions on Intelligent Transportation Systems*, 17(10):2681–2698, 2016.
116. Xingchen Zhang, Yuxiang Feng, Panagiotis Angeloudis, and Yiannis Demiris. Monocular visual traffic surveillance: a review. *IEEE Transactions on Intelligent Transportation Systems*, 23(9):14148–14165, 2022.
117. Nicolas Saunier, Tarek Sayed, and Karim Ismail. Large-scale automated analysis of vehicle interactions and collisions. *Transportation Research Record*, 2147(1):42–50, 2010.
118. Karim Ismail. Application of computer vision techniques for automated road safety analysis and traffic data collection. *Ph.D. thesis, University of British Columbia,* 2010.
119. Joshua Stipancic, Luis Miranda-Moreno, Nicolas Saunier, and AurAlie Labbe. Network screening for large urban road networks: Using GPS data and surrogate measures to model crash frequency and severity. *Accident Analysis and Prevention*, 125:290–301, 2019.
120. Paul St-Aubin, Nicolas Saunier, Luis Miranda-Moreno, and Karim Ismail. Use of Computer Vision Data for Detailed Driver Behavior Analysis and Trajectory Interpretation at Roundabouts. *In Transportation Research Record: Journal of the Transportation Research Board,* 2389(1), 2013.
121. Nopadon Kronprasert, Chomphunut Sutheerakul, Thaned Satiennam, and Paramet Luathep. Intersection safety assessment using video-based traffic conflict analysis: The case study of Thailand. *Sustainability*, 13(22):12722, 2021.
122. Alexey Bochkovskiy, Chien-Yao Wang, and Hong-Yuan Mark Liao. YOLOv4: Optimal speed and accuracy of object detection. *CoRR* abs/2004.10934, 2020.
123. Simon N. Wood. Thin plate regression splines. *Journal of the Royal Statistical Society: Series B (Statistical Methodology)*, 65(1):95–114, 2003.
124. Terrell Nathan Mundhenk, Michael J. Rivett, Xiaoqun Liao, and Ernest L. Hall. Techniques for fisheye lens calibration using a minimal number of measurements. In David P. Casasent, editor, *Intelligent Robots and Computer Vision XIX: Algorithms, Techniques, and Active Vision*. volume 4197, pp. 181–190, International Society for Optics and Photonics. SPIE, 2000.
125. Qishen Zhou, Roozbeh Mohammadi, Weiming Zhao, Kaihang Zhang, Lihui Zhang, Yibing Wang, Claudio Roncoli, and Simon Hu. Queue profile identification at signalized intersections with high-resolution data from drones. In *2021 7th International Conference on Models and Technologies for Intelligent Transportation Systems (MT-ITS)*, pp. 1–6, 2021:.
126. FDOT. Signal Operating Procedure. `https://www.fdot.gov/docs/default-source/roadway/DS/13/IDx/17870.pdf`.
127. FDOT. Safety. `https://safety.fhwa.dot.gov`.
128. Eun-Ha Choi. *Crash factors in intersection-related crashes: An on-scene perspective*. Technical report National Highway Traffic Safety Admin., Washington, DC, 2010.
129. A. Houenou, P. Bonnifait, V. Cherfaoui, and W. Yao. Vehicle trajectory prediction based on motion model and maneuver recognition. In *2013 IEEE/RSJ International Conference on Intelligent Robots and Systems*, pp. 4363–4369, 2013.
130. John M. Scanlon, Rini Sherony, and Hampton C. Gabler. Injury mitigation estimates for an intersection driver assistance system in straight crossing path crashes in the United States. *Traffic Injury Prevention*, 18(Suppl. 1):S9–S17, 2017.

131. Xin Huang, Stephen G McGill, Brian C Williams, Luke Fletcher, and Guy Rosman. Uncertainty-aware driver trajectory prediction at urban intersections. In *2019 International Conference on Robotics and Automation (ICRA)*, pp. 9718–9724, IEEE, 2019.
132. John M. Scanlon, Rini Sherony, and Hampton C. Gabler. Preliminary potential crash prevention estimates for an Intersection advanced driver assistance system in straight crossing path crashes. In *2016 IEEE Intelligent Vehicles Symposium (IV)*, pp. 1135–1140, IEEE, 2016.
133. Stéphanie Lefèvre, Dizan Vasquez, and Christian Laugier. A survey on motion prediction and risk assessment for intelligent vehicles. *ROBOMECH Journal*, 1(1):1–14, 2014.
134. Dizan Vasquez and Thierry Fraichard. Motion prediction for moving objects: A statistical approach. In *Proceedings: IEEE International Conference on Robotics and Automation, 2004*, volumn 4, pp. 3931–3936, IEEE, 2004.
135. Joshua Joseph, Finale Doshi-Velez, Albert S Huang, and Nicholas Roy. A Bayesian nonparametric approach to modeling motion patterns. *Autonomous Robots*, 31(4):383, 2011.
136. Quan Tran and Jonas Firl. Online maneuver recognition and multimodal trajectory prediction for intersection assistance using non-parametric regression. In *2014 IEEE Intelligent Vehicles Symposium Proceedings*, pp. 918–923, IEEE, 2014.
137. Sepideh Afkhami Goli, Behrouz H Far, and Abraham O Fapojuwo. Vehicle trajectory prediction with gaussian process regression in connected vehicle environment. In *2018 IEEE Intelligent Vehicles Symposium (IV)*, pp. 550–555, IEEE, 2018.
138. Alexandre Alahi, Kratarth Goel, Vignesh Ramanathan, Alexandre Robicquet, Li Fei-Fei, and Silvio Savarese. Social LSTM: Human trajectory prediction in crowded spaces. In *Proceedings of the IEEE Conference on Computer Vision and Pattern Recognition*, pp. 961–971, 2016.
139. ByeoungDo Kim, Chang Mook Kang, Jaekyum Kim, Seung Hi Lee, Chung Choo Chung, and Jun Won Choi. Probabilistic vehicle trajectory prediction over occupancy grid map via recurrent neural network. In *2017 IEEE 20th International Conference on Intelligent Transportation Systems (ITSC)*, pp. 399–404, IEEE, 2017.
140. Nachiket Deo and Mohan M Trivedi. Convolutional social pooling for vehicle trajectory prediction. In *Proceedings of the IEEE Conference on Computer Vision and Pattern Recognition Workshops*, pp. 1468–1476, 2018.
141. Seong Hyeon Park, ByeongDo Kim, Chang Mook Kang, Chung Choo Chung, and Jun Won Choi. Sequence-to-sequence prediction of vehicle trajectory via LSTM encoder–decoder architecture. In *2018 IEEE Intelligent Vehicles Symposium (IV)*, pp. 1672–1678, IEEE, 2018.
142. Tianyang Zhao, Yifei Xu, Mathew Monfort, Wongun Choi, Chris Baker, Yibiao Zhao, Yizhou Wang, and Ying Nian Wu. Multi-agent tensor fusion for contextual trajectory prediction. In *Proceedings of the IEEE Conference on Computer Vision and Pattern Recognition*, pp. 12126–12134, 2019.
143. Yihan Cai, Menghan Tian, Weidong Yang, and Yi Zhang. Stay point analysis in automatic identification system trajectory data. In *Proceedings of the 2018 International Conference on Data Science*, pp. 273–278, 2018.
144. T. Banerjee, X. Huang, K. Chen, A. Rangarajan, and S. Ranka. Clustering object trajectories for intersection traffic analysis. In *6th International Conference on Vehicle Technology and Intelligent Transport Systems (VEHITS)*, 2020.
145. Hongling Wang, Joseph Kearney, and Kendall Atkinson. Arc-length parameterized spline curves for real-time simulation. In *Proceedings of the 5th International Conference on Curves and Surfaces*, volumn 387396, 2002.

146. Kichun Jo, Minchul Lee, Junsoo Kim, and Myoungho Sunwoo. Tracking and behavior reasoning of moving vehicles based on roadway geometry constraints. *IEEE Transactions on Intelligent Transportation Systems*, 18(2):460–476, 2016.
147. Yulai Wan and Anming Zhang. Urban road congestion and seaport competition. *Journal of Transport Economics and Policy*, 47(1):55–70, 2013.
148. Henry X Liu and Wenteng Ma. *Virtual probe approach for time-dependent arterial travel time estimation*. Technical report, Transportation Research Board, 2008.
149. Jiuqing Wan and Liu Li. Distributed optimization for global data association in non-overlapping camera networks. In *2013 Seventh International Conference on Distributed Smart Cameras (ICDSC)*, pp. 1–7, IEEE, 2013.
150. Shu Zhang, Elliot Staudt, Tim Faltemier, and Amit K. Roy-Chowdhury. A camera network tracking (CamNeT) dataset and performance baseline. In *2015 IEEE Winter Conference on Applications of Computer Vision*, pp. 365–372, IEEE, 2015.
151. Kuan-Wen Chen, Chih-Chuan Lai, Pei-Jyun Lee, Chu-Song Chen, and Yi-Ping Hung. Adaptive learning for target tracking and true linking discovering across multiple non-overlapping cameras. *IEEE Transactions on Multimedia*, 13(4):625–638, 2011.
152. Andrew Gilbert and Richard Bowden. Tracking objects across cameras by incrementally learning inter-camera colour calibration and patterns of activity. In *European Conference on Computer Vision*, pp. 125–136, Springer, 2006.
153. Cheng-Hao Kuo, Chang Huang, and Ram Nevatia. Inter-camera association of multi-target tracks by on-line learned appearance affinity models. In *European Conference on Computer Vision*, pp. 383–396, Springer, 2010.
154. Dimitrios Makris, Tim Ellis, and James Black. Bridging the gaps between cameras. In *Proceedings of the 2004 IEEE Computer Society Conference on Computer Vision and Pattern Recognition, 2004*, volumn 2, pp. II-205–II-210. IEEE, 2004.
155. Robert L. Gordon. *Traffic Signal Retiming Practices in the United States*, volume 409. Transportation Research Board, 2010.
156. Aude Hofleitner, Ryan Herring, and Alexandre Bayen. Arterial travel time forecast with streaming data: A hybrid approach of flow modeling and machine learning. *Transportation Research Part B: Methodological*, 46(9):1097–1122, 2012.
157. Ruimin Li and Geoffrey Rose. Incorporating uncertainty into short-term travel time predictions. *Transportation Research Part C: Emerging Technologies*, 19(6):1006–1018, 2011.
158. Ashish Bhaskar, Edward Chung, and André-Gilles Dumont. Fusing loop detector and probe vehicle data to estimate travel time statistics on signalized urban networks. *Computer-Aided Civil and Infrastructure Engineering*, 26(6):433–450, 2011.
159. Filmon G. Habtemichael and Mecit Cetin. Short-term traffic flow rate forecasting based on identifying similar traffic patterns. *Transportation Research Part C: Emerging Technologies*, 66:61–78, 2016.
160. D Nikovski, N Nishiuma, Y Goto, and H Kumazawa. Univariate short-term prediction of road travel times. In *Proceedings: 2005 IEEE Intelligent Transportation Systems*, pp. 1074–1079, IEEE, 2005.
161. Shu-Kai S Fan, Chuan-Jun Su, Han-Tang Nien, Pei-Fang Tsai, and Chen-Yang Cheng. Using machine learning and big data approaches to predict travel time based on historical and real-time data from Taiwan electronic toll collection. *Soft Computing*, 22(17):5707–5718, 2018.

162. Mahmood Rahmani, Erik Jenelius, and Haris N Koutsopoulos. Non-parametric estimation of route travel time distributions from low-frequency floating car data. *Transportation Research Part C: Emerging Technologies*, 58:343–362, 2015.
163. Andreas Allström, Joakim Ekström, David Gundlegård, Rasmus Ringdahl, Clas Rydergren, Alexandre M Bayen, and Anthony D Patire. Hybrid approach for short-term traffic state and travel time prediction on highways. *Transportation Research Record*, 2554(1):60–68, 2016.
164. Xianyuan Zhan, Satish V. Ukkusuri, and Chao Yang. A Bayesian mixture model for short-term average link travel time estimation using large-scale limited information trip-based data. *Automation in Construction*, 72:237–246, 2016.
165. Xinchen Liu, Wu Liu, Tao Mei, and Huadong Ma. Provid: Progressive and multimodal vehicle reidentification for large-scale urban surveillance. *IEEE Transactions on Multimedia*, 20(3):645–658, 2017.
166. Zhedong Zheng, Liang Zheng, and Yi Yang. A discriminatively learned CNN embedding for person reidentification. *ACM Transactions on Multimedia Computing, Communications, and Applications (TOMM)*, 14(1):1–20, 2017.
167. J. Sha, Y. Zhao, W. Xu, H. Zhao, J. Cui, and H. Zha. Trajectory analysis of moving objects at intersection based on laser-data. In *ITSC*, pp. 289–294, 2011.
168. H. Xu, Y. Zhou, W. Lin, and H. Zha. Unsupervised trajectory clustering via adaptive multi-kernel-based shrinkage. In *2015 IEEE International Conference on Computer Vision (ICCV)*, pp. 4328–4336, 2015.
169. S. Al-Dohuki, F. Kamw, Y. Zhao, X. Ye, and J. Yang. Trajanalytics: An open source geographical trajectory data visualization software. In *The 22nd IEEE Intelligent Transportation Systems Conference*, 2019.
170. Wooil Kim, Changbeom Shim, Ilhyun Suh, and Y. Chung. A visual explorer for analyzing trajectory patterns. *IEEE Conference on Visual Analytics Science and Technology (VAST)*, pp. 199–200, 10 2017.
171. Stan Salvador and Philip Chan. FastDTW: Toward accurate dynamic time warping in linear time and space. In *KDD Workshop on Mining Temporal and Sequential Data*, Citeseer, 2004.
172. Tania Banerjee, Yashaswi Karnati, Ke Chen, Yury Lebedev, Aotian Wu, Rahul Sengupta, Anand Rangarajan, and Sanjay Ranka. A multi-sensor edge-based system for intelligent traffic intersections. In *2022 Fifth International Conference on Connected and Autonomous Driving (MetroCAD)*, pp. 39–46, 2022.
173. Sanjay Ranka, Anand Rangarajan, Lily Elefteriadou, Siva Srinivasan, Emmanuel Poasadas, Dan Hoffman, Raj Ponnulari, Jeremy Dilmore, and Tom Byron. A vision of smart traffic infrastructure for traditional, connected, and autonomous vehicles. In *2020 International Conference on Connected and Autonomous Driving (MetroCAD)*, pp. 1–8, 2020.
174. Ahan Mishra, Ke Chen, Subhadipto Poddar, Emmanuel Posadas, Anand Rangarajan, and Sanjay Ranka. Using video analytics to improve traffic intersection safety and performance. *Vehicles*, 4(4):1288–1313, 2022.
175. Aotian Wu, Tania Banerjee, Anand Rangarajan, and Sanjay Ranka. Trajectory prediction via learning motion cluster patterns in curvilinear coordinates. In *ITSC 2021, Indianapolis, IN, USA, September 19–22, 2021*, pp. 2200–2207, IEEE, 2021.
176. Xiaohui Huang, Pan He, Anand Rangarajan, and Sanjay Ranka. Machine-learning-based real-time multi-camera vehicle tracking and travel-time estimation. *Journal of Imaging*, 8(4):101, 2022.

Index